Global Climate Change, the Bible, & Science

Dennis R. Dinger

Global Climate Change, the Bible, & Science

All Bible verses taken from the KJV.

ISBN 978-0-557-80440-5

Contents

Preface

Each of us is a product of our parents, our God-given talents, our environment, and the sum total of our life-long intellectual activities. The books we read, the subjects we study, the discussions and arguments we join, our chosen fields of employment, etc., all contribute to our way of thinking, our attitudes, and especially to our opinions. Each of us is unique in that way.

As a Christian, I have definite religious beliefs. As an Emeritus Professor of Ceramic and Materials Engineering, and having been a Licensed Professional Engineer, a Ceramic Engineer, a Computer Programming Instructor, and a Combustion Engineer, I have studied, applied, and taught a wide variety of technical scientific and engineering phenomena. Throughout my life, I have also read, studied, discussed, and contemplated a wide variety of books and ideas in the realms of art, religion, and politics. All of these experiences have molded the way I think and the way I address new subjects.

As a research scientist and computer programmer, I have written and guided the development of several computer models. I know, for example, that I could write a model that could prove practically **anything** by judiciously picking and choosing the phenomena to include in the model and to carefully control the algorithms used for their simulation. Although such an intentionally erroneous model would not be legitimate, I believe I have the experience needed to easily implement one.

Bad models and bad results happen all the time — not because someone sets out to intentionally ignore important phenomena, but because while attempting to make a good model, they have unknowingly overlooked applicable phenomena that are necessary to the success of their model. Bad models are also produced when accidental or purposeful results happen to agree with preconceived notions.

I have seen erroneous models first hand in my chosen field of study — rheology (the science of fluid flow.) I know for a fact that if a model does not contain code to simulate pertinent phenomena, the results **cannot** predict reality. Results may look accurate and the model might even seem to be useful. But without appropriate algorithms, they cannot be accurate. Severe criticism can also come when necessary phenomena **are** included in models that the rest of the scientific community deems unnecessary. It all depends on who is creating or critiquing the model and on his or her understanding of the phenomena being modeled.

I know from experience that computer models are not valid **until tested against real data**. Each of my models had to be tested against real laboratory or industrial systems before anyone would believe the results.

With regard to the topics of this book, I know for a fact that meteorologists **cannot** accurately predict today's weather, yet we are to believe that their cousins — the climatologists — have developed models that can accurately predict what will happen to our climate 100 years from now. If today's climate models are so accurate, our predictions for today's weather should be 100% certain all the time. But they are not. The current weather models **cannot** do it. Why, then, should we put any faith in climate change models? I think the answer is: We should NOT believe them! We have an obligation to question them!

Many present and former politicians use their soap boxes to tell us (the public) what we should and shouldn't be thinking on technical issues such as global warming. They **believe** computer model results are totally accurate (without necessarily knowing anything about computer models or about anything technical.) They **want to think** computer model results are totally accurate because those results fit their agendas and their preconceived answers. And once they reach that point, they rush to stifle further debate, to make decisions (the Supremes), and to pass laws (Congress and the President) that will affect us, our children, our grandchildren, etc., for many decades to come.

The alarmist global warming community has attempted to limit debate to only those few who have studied climate change as their professional specialty. The politicians, who are part of this alarmist community, are not qualified (by their own criteria) to voice any opinions on the subject — but that doesn't stop them. They are **important** leaders and we average citizens are supposed to bow and kowtow to their superior knowledge, opinions, and conclusions. Hogwash!!

Have any of these lawmakers and politicians ever formally studied climate change? Have any of them ever studied engineering or science? Are any of them the slightest bit technically literate? Do scientific discussions really end when **consensus** has been achieved among scientists? Is anthropogenic global warming a real threat to society because computer models say so, or because power-hungry lawmakers, the political left, and the media say so? Has catastrophic global warming really been occurring? These questions can all be answered with a resounding, "**NO!!!**"

So many questions remain unanswered regarding the phenomenon known as *global climate change*!! Where are the answers? What scientific phenomena should modelers be considering that they have not? We will address such questions throughout this book.

As a scientist, I have read books and articles by scientists which were extremely difficult to understand. In many cases, the language and logic were so high-brow, the authors' arguments and trains-of-thought were difficult to follow. The cause for some of this obfuscation was to allow scientists to tiptoe around the **religion versus science** debate. It appears some want to make religious arguments, but they dare not — because the opposition would label their arguments religion, rather than science, and dismiss them out of hand.

Many of those trying to prove that intelligent design (ID) is the method of creation have done this. They want to prove, by scientific methods alone, that intelligence was required in the designs of the universe, our solar system, and our planet and all life on it. They want to do so without every mentioning God. God forbid they should ever mention "God" in their scientific papers! Their logic is both complex and precise to the extreme. The result: they produce papers that are difficult to read and to understand.

I don't feel an obligation to tiptoe around the subject of God without mentioning the three letter word — G-O-D. I am a Christian and I believe the Bible. **Just because the Bible is a religious book doesn't automatically invalidate everything it says.** There is a lot of good information in the Bible, if only mankind would look into it. I plan to do that in this book.

I also plan to discuss scientific issues pertinent to the global warming debate in a way that the average person can understand. Some critical issues, looked at from the proper points-of-view, don't appear critical at all. From some points-of-view, those same issues may carry extreme importance, but for different reasons than are heard in today's public discussions. I will do my best to write clearly and simply so everyone can understand.

Regarding the climate change debate, the Bible, and current science, this book records my questions, answers, views, and opinions, and emphasizes the issues that I deem to be important.

May my thoughts and contributions re-ignite thoughtful public debate.

Dennis R. Dinger
25 June 2010

1

Introduction

Currently ...

21st Century earth. July, 2010. United States of America. Barack Obama is President. The Health Care bill is law. Crude oil spills into the Gulf of Mexico. The Climate Bill, known as Cap & Trade, is supposedly next on the agenda. Our southern border is not secure. Global warming continues to be a worldwide concern. Politics rules the day.

Every argument on every topic is twisted by politicians and the news organizations who cover them. Polls show that the majority of Americans didn't want universal health care. It passed anyway. The administration says it is on top of the oil spill. But they don't know how to stop it, nor do they have the equipment or expertise to stop it! The Justice Department is suing the State of Arizona over their new law regarding illegal aliens. Our southern border is full of holes and the feds (and the President) don't appear to care.

What choices do we citizens have? We may not have any choices until election days in November 2010 and 2012. Those days seem like eternities away.

Important Topics

Let's consider some important topics that should be under discussion by all Americans. Are they being discussed publicly? Are we going to have any choices in these matters?

Global Warming

Global satellite temperature data shows there has been no global warming this decade (2000-2010). During this time, the average global temperature has remained essentially constant. Have you heard that reported in the news? I haven't. In the mass media, almost every weather event is blamed on global warming. There has been no global warming for ten or more years, yet every new weather system is blamed on global warming. That makes no sense (unless there is a hidden agenda at stake)!

Unless you are a climate change scientist, *the intelligentsia* says your opinion on global warming doesn't count. But *the intelligentsia*, which includes politicians, public figures, the media, actors, actresses, and anyone else who considers themselves to be important (none of whom appear to be climate change scientists), are saying that global warming continues despite data that show otherwise. And they say it as if it's totally obvious to everyone — only morons wouldn't recognize it. "Did global warming cause yesterday's heat wave?" "Of course it did!," they tell us. And so it goes.

What about debate on this subject? The environmental alarmists, and the media who report their every word, want us to believe the debate is over. **Consensus**, we are told, has been established. We the people are the cause of the warming. And with consensus, the debate supposedly ended. I don't remember the debate ever having begun, but we are led to believe it ended already.

Where is our voice in this matter? We don't have one — and we won't have one if the alarmists have anything to say about it.

Only Facts Allowed in the Classroom

Darwinian evolution is taught as "factual science" in public schools. This occurs despite recent studies that show that life in single-celled creatures[1] is much more complex ("irreducibly complex") than anyone in Darwin's time anticipated. Creationism, which relies on religious belief, miracles, and God, is treated as a joke — we certainly can't teach that in the classroom!

---

[1]Behe, Michael J., <u>Darwin's Black Box</u>, Free Press, New York, 1996, 2006.

Intelligent Design[2] (ID) is considered to be a form of creationism so it can't be taught either. The courts declared it unconstitutional to teach ID in the public schools — "separation of church and state," you know! Yet evolution is also a religion.[3] Why? Because no one has proven it (nor can they). It requires lots of faith.

Sure — evolution within each species occurs. Dogs are large, small, hairy, hairless, brown, white, black, etc. The wide range of variations we see in dogs are all possibilities within the canine genome. **Evolution** (*microevolution*, that is) **exists**. Yes — microevolution, which enables 'survival of the fittest' and variability — exists.

But the capability of a dog to evolve into a cat or a pony or a lizard — or the capability of a monkey to evolve into a human — which is known as *macroevolution* — has never been proven nor demonstrated. Man could not have evolved from a single-celled creature without macroevolution, and evolutionists cannot demonstrate that form of evolution. They can draw great pictures, make great videos, offer detailed explanations, but can they find real evidence that demonstrates it? No.

Despite claims of "overwhelming evidence" in support of **evolution**, the evidence supports only **microevolution** — not **macroevolution**. This is a huge word game — an intentional deception — that evolutionists are playing.

This needs repeating. *Microevolution* is being used as proof of *macroevolution*. Both are known collectively as ***evolution***. They are not the same. Microevolution allows for all the variety within each species. Microevolution exists. Until they can demonstrate it, however, macroevolution remains only a theory. Macroevolution is the form of evolution that allows lesser species to evolve into more complex species. Regarding macroevolutionary arguments, the biochemist Behe[4] wrote, "Persuasive evidence to support that position, however, has not been forthcoming."

---

[2] Kitzmiller v. Dover Area School District, Wikipedia, Kitzmiller Tammy, et al. v. Dover Area School District, et al. (400 F. Supp. 2d 707, Docket no. 4cv2688).

[3] http://www.creationists.org/evolutionism-is-a-religion.html

[4] Behe, *op.cit.*, p.15.

Without solid, supportive evidence, to believe in macroevolution requires faith — lots of it! That makes macroevolution (or just simply ***evolution***) a religion. You won't see much use of the prefixes micro- and macro- in regard to the word *evolution*. Supporters just use the term ***evolution***. They lump it all into that one term. But the form of evolution necessary to demonstrate their argument that all of today's complex life evolved from single-celled species (macroevolution) has never been demonstrated.

For this reason, evolutionary theory should be unconstitutional, too. It is a religion. But if they banned it, they would have no way to explain how we all came into existence. And evolutionary theory has been taught **as fact** for years and years in the public schools. The vast majority of Americans have public school educations where they learned all about evolution. They were all taught that we evolved from lower life forms. Most, therefore, believe that ***evolution*** is both science and fact.

Some Americans still want to correct the record and put ID and creationism back into the classrooms as alternative theories of creation. But when a school board tried to require discussion of intelligent design alongside evolution in biology classes in the Dover Area School District in Pennsylvania, several families sued. The verdict: ID is a form of creationism, the teaching of which is unconstitutional. This case showed that there are many people out there who like the status quo and who are willing to defend it in the courts.

The issue has not gone away, however. It still hangs around. Hardly anyone wants to talk about it anymore, but it will continue to surface from time to time. Many are tired of the debate. They want it to end. And the side that is in power — the evolutionists — would be happy to never need to debate this issue again.

Do we have a choice here? If you don't believe in evolution, you can always home-school your children or send them to a Christian school. That's your choice. It is preferable, and less expensive, than long, drawn-out court proceedings.

Uniformitarianism vs Catastrophism

Many believe the earth was created gradually over millions of years. This is known as *gradualism* or *uniformitarianism*. This theory says that all changes today happen the same way they happened millions of years ago — and all changes happen very, very slowly. The term *uniformitarianism* was coined in the early 19th Century and it has had many supporters throughout the 19th and 20th Centuries.

The alternative creation view is known as *catastrophism*. It says that the earth was formed by a series of "sudden, short-lived, violent events, possibly worldwide in scope."[5] The Rocky Mountains and the Grand Canyon, for example, took millennia to achieve today's shape and appearance according to uniformitarianism, but they only took a very short period of time to form according to catastrophism.

Catastrophism allows for a much younger earth than is commonly taught by uniformitarianism. Since a young earth is consistent with the Biblical explanation of creation, catastrophism has generally been discredited for its close ties to creationism.

Biblical *creationism* was the prevailing scientific view up until the early 19th Century.[5] But when uniformitarianism was defined, it displaced creationism as the prevailing view. The uniformitarian theory that earth, mankind, animals, etc., required millions of years to slowly evolve into their current states, doesn't agree with the biblical explanation. Furthermore, mankind has seen no major worldwide catastrophes in recent history — so uniformitarianism is supported by mankind's experience.

Many who support uniformitarian principles are Darwinian evolutionists. Uniformitarianism and Darwinian evolution fit well together. Each requires millions of years and numerous small changes for full development to occur. In fact, in his 1955 book[6], Immanuel Velikovsky, a proponent of the catastrophist theory of creation, says that evolution grew out of uniformitarianism. "One after the other, scenes of upheaval and devastation have presented themselves to explorers, and almost every new cave opened, mountain thrust explored, undersea canyon investigated, has consistently disclosed the same picture of violence and desolation. **Under the weight of this evidence two great theories of the nineteenth century have become more and more strained: the theory of uniformity and the theory of evolution built upon it."** (emphasis added)

---

[5] Wikipedia, "Catastrophism."

[6] Velikovsky, Immanuel, Earth in Upheaval, 1955, p107.

How many of you readers have ever heard of Velikovsky, his books, or his ideas? Why are all of the various theories of creation and Velikovsky's ideas not being openly debated in the public arena? Is it because the scientific community of the mid-20th Century tried to stop the publication of Velikovsky's ideas — hoping those ideas would never see the light of day? (Despite their efforts, Velikovsky's books were published.) Is it because the scientific community knows catastrophism is correct and uniformitarian evolution is not defendable? Is it because the public doesn't want to believe the truth about evolutionary theory? ... that it's not a science, but a faith? Is it because the public is worried evolutionary theory might not be correct at all? Is it that the mass media wants everyone to believe the arguments for evolution, so it isn't faithfully reporting opposing views? Is it that many people today want nothing to do with God or any theories that rely on God? Is it that many are clinging to evolutionary theory because they don't like the alternative? ... which means they would have to deal with the question, "Is there really a God of creation?" Each of these proposals is true for some people.

Do we have any choices here? We can discuss the various creation theories and ask questions — but not in public schools. Can we discuss creation openly in society? Dissenters are doing their best to stop new discussions.

Pollutant Gases

Which gases are pollutant gases according to the US government? Ozone and carbon dioxide are. Ozone was labeled a pollutant gas many years ago by the EPA. It is a very useful and helpful gas, but it carries the label as a pollutant anyway. Hardly anyone knows it is beneficial, compliments of the EPA.

Recently, carbon dioxide was also defined as a pollutant by the US Supreme Court. Since carbon dioxide is integral to life on this planet, it is strange that anyone would label it a pollutant. But stranger things have happened in Washington, DC.

Ozone

Ozone (O_3) is a very reactive form of gaseous oxygen. In contains three atoms of oxygen, whereas normal oxygen molecules only contain two atoms of oxygen. The first two atoms of oxygen in ozone are tightly bonded as they are in normal oxygen, but the third oxygen atom is loosely bonded so it reacts easily — even at room temperature.

Ozone oxidizes and removes many pollutants from the air we breathe. It does all this at typical daily temperatures. Because it helps keep our atmosphere clean, we are fortunate that ozone exists.

By comparison, normal oxygen molecules (O_2) are very stable and cannot easily oxidize pollutants at room temperatures. Normal oxygen has to be heated to higher temperatures before it can react with and oxidize hydrocarbons and other pollutants.

Ozone forms naturally near waterfalls, during lightning storms, and when UV solar radiation reacts with oxygen molecules. Without ozone in our atmosphere, pollutants **would** pile up, and then we would **really** be in a pickle with only polluted, nasty air to breathe. But because ozone has been declared a pollutant gas, most people believe it to be bad and hazardous to health. At low concentrations, it is quite beneficial. To not have ozone at all in our atmosphere would be hazardous to everyone's health.

One reason we receive daily ozone alerts to learn when our air is going to be especially polluted is that its rate of formation increases as levels of atmospheric pollution increase. On days when smog is especially concentrated, lots of ozone is created as the sun's UV rays are scattered off the numerous pollutant particles. Smog contains thousands of different molecules and particles which are intermediate combustion products of hydrocarbon fuels. Each of these are present in such low concentrations in smog that it is difficult to identify any one and measure its individual concentration. But since the rate of formation of ozone and its concentration increases as smog increases, ozone is the molecule present in greatest abundance and the easiest to analyze. So ozone is monitored, and that gives us "ozone alerts."

Ozone is a natural scavenger molecule. The more smog present, the more ozone will form to clean up the smog. Ozone is a very useful, naturally forming molecule that we have been led to believe is a pollutant gas. It really is not a pollutant at all.

And mysteriously, it just happens to form in needed quantities. When pollution is low, little ozone forms. When pollution is high, lots of ozone forms. It forms in just the right amounts to function properly as the scavenger molecule it is. Sounds like a natural cycle and a natural balance to me!

Carbon Dioxide

Then there is carbon dioxide. Carbon dioxide (CO_2) has recently (April 2007) been declared a pollutant gas by the US Supreme Court. Carbon dioxide is a naturally formed gas that participates in all life on this planet. For that reason alone, it is crazy to label it a *pollutant*.

The liberal wing of the court decided 5 to 4 that carbon dioxide is a problem.[7] I submit that their ruling oversteps their mandate. I agree with the dissenting opinions of the case.[8] Based on the court's ruling, nevertheless, the EPA is now supposed to monitor and regulate CO_2 concentrations and industrial emissions flowing into our atmosphere. Why?

Apparently, because bogus science has declared carbon dioxide to be the main cause of anthropogenic global warming. It wasn't clear at the time whether global warming even truly existed. And now we know it hasn't existed for the last 15 years. It certainly hasn't been proven that carbon dioxide is the cause of this non-existent dire problem. How can you identify the specific cause for a problem that doesn't even exist? No one ever said politics has to be logical.

Three years after the Supreme Court decision, the former Director of the University of East Anglia's Climate Research Unit (which was at the center of this controversy) admitted that there has been no "statistically significant" global warming over the last 15 years (1995-2010).[9] I submit that the proceedings and decision by the US Supreme Court in 2007 were a bit premature. But the decision has been made. We will see if it is ever rescinded.

We need carbon dioxide for all life on this planet. For it to be declared a pollutant gas that needs to be regulated is sheer lunacy.

When and where in society today are we discussing ozone? How many people even know what ozone is, how great an oxidizer it is, and how

---

[7] Justice Stevens, "Massachusetts v EPA," US Supreme Court, No. 05-1120, 2 Apr 2007, Opinion of the Court.

[8] Chief Justice Roberts, and Justice Scalia, "Massachusetts v EPA," US Supreme Court, No. 05-1120, 2 Apr 2007, Dissenting Opinions.

[9] Petre, Jonathan, UK Daily Mail: "Climategate U-turn as Scientist at Centre of Row Admits: There has Been No Global Warming Since 1995," 11 Aug 2010.

wonderfully it cleans up our atmosphere? Why aren't we discussing it publicly? Why are all good topics for discussion being hidden from us?

When and where in society today are we discussing carbon dioxide? Man and animal alike exhale carbon dioxide. Plants use it for photosynthesis. But now that court decisions have been made, we are left with the consequences.

Computer Models

We are told by the intelligentsia that the US and the world are in imminent peril from anthropogenic global warming (i.e., global warming **caused by mankind.**) Why? Because computer models have predicted it. Meteorologists cannot accurately predict today's weather, yet in theory, the climatologists can accurately tell us what will happen hundreds of years from now. Well, I don't think so!! It's all a bunch of baloney!

It is doubtful that most people know anything about computer models. We will consider computer models in much more detail in a later chapter.

If you are familiar with computer models, and you believe those that predict catastrophic climate change are as accurate as they claim, I have some swampland I'd like to sell you. But see ... it doesn't really matter what you or I think or whether the arguments about computer models even make sense. The *big boys* have decided. That's it! The discussion is over. **Consensus** has been achieved.

Who are we to object? To borrow a characterization from BP Chairman Carl-Henric Svanberg[10] of 2010 Gulf Oil Spill fame, who are we, "the small people," to object?

What do we know? Nothing much, apparently.

God

In America, society and politics are wanting us to believe that God is irrelevant. In fact, the Christian God, any mention of Him, and anything at all to do with Him, is being erased from schools, open Christmas displays, reports,

---

[10] Speaking in Washington on 16 June 2010, "We care about the small people."

and discussions throughout the land — as we speak! Any mention of Him, according to political correctness, **ist verboten!**

Christianity is openly mocked. Christians are ridiculed. The Bible was recently characterized by an outspoken liberal as one of several "works of fiction" for a "child-like audience."[11]

Speak out against Mohammed or Islam and your life may be in jeopardy. Speak out for Mohammed and Islam and that is fine. Anything to do with any non-Christian religion is apparently OK, too — society is quite tolerant of all *other* religions. Speak for Christ or Christianity and you are labeled a kook and disregarded. Glenn Beck has been trying to make the case on his radio and TV programs that God is not only very relevant to today's discussions — He is necessary! But Beck has been labeled a crazy man[12] whom citizens are encouraged to ignore.

Some Moslems want to build an Islamic Center near Ground Zero in New York City. Apparently, they have already overcome lots of hurdles from the city to begin construction. The Greek Orthodox church that was demolished when one of the twin towers fell on it has not yet been given permission to rebuild. One is Islamic. One is Christian. Notice which one is progressing quickly — despite mounting protests.

Try to suggest the addition of creationism to public school curricula, and you are unconstitutional[13]. Label evolution as the religious argument it really is[14-15], and you might suffer the wrath of society. So much for freedom of speech, freedom of religion, and Christianity in the 21st Century. Apparently,

---

[11] Garofalo, Janeane, 24 June 2010, interview with the Huffington Post.

[12] http://www.huffingtonpost.com/bob-cesca/glenn-beck-and-the-conseq_b_184936.html, Cesca, Bob, 9 Apr 09, "Glenn Beck and the Consequences of Crazy Talk"

[13] Kitzmiller, *op.cit.*

[14] http://www.creationists.org/evolutionism-is-a-religion.html

[15] Lovgren, Stefan, "Evolution and Religion Can Coexist, Scientists Say," http://news.nationalgeographic.com

some religions are more politically correct and more acceptable than others. Christianity is certainly not on the list of PC religions.

What choice do we have? To date, we Americans are still free to worship the god of our choice. The God of creation is worshiped by many. But say His name openly in the wrong public place and you may find yourself in trouble with the PC police.[16]

PC Thought

Political correctness doesn't just stop with climate discussions and religion. It extends into the sciences.

Peer-reviewed scientific journals are the only authorized places for scientists and engineers to publish. Many academics and those working on research projects are expected to publish as a requirement for promotion. Peer-reviewed publications count very highly. Non-peer reviewed publications don't count much at all. If the research is of any value at all, it must be published in a peer-reviewed journal.

But many editors and reviewers of peer-reviewed journals require politically correct (PC) opinions and comments. Try to insert thoughts that run against the PC grain and the paper will not be published nor see the light of day. Try this too many times and your career might be ruined. This is a form of censorship masquerading as a legitimate, highly esteemed process. Is this process about to change? It certainly doesn't appear so.

What can be done about this? Not much. To obtain approval for the publication of some papers, authors have to change so much to meet reviewers' requirements that the finished paper no longer makes the author's original point. For this reason, many authors are choosing to publish books elsewhere — for example, in the open press. They are taking their scholarly work directly to the people — to those who will read and benefit from their efforts.

Which authors should you read? That is your choice — select wisely.

[16] "Cops: Kids Can't Pray at U.S. Supreme Court, Silenced! Christian students ordered to stop devotional on public grounds," 29 Aug 2010, World Net Daily, www.wnd.com.

Censorship

We are supposedly an open society with freedom of the press, but we nevertheless must still deal with forms of censorship. The preceding example showed a form of censorship. But more direct forms of censorship are waiting to rear their ugly heads — even in our open society.

For example, Velikovsky's first book, Worlds in Collision, 1950, was considered to be so radical by the scientific elite of the day that they tried to prevent its publication. Velikovsky supported catastrophism at a time when the only approach acceptable to the scientific elite was uniformitarianism. And so they tried to prevent the publication of that book. They didn't succeed, but they caused a big brouhaha that surrounded Velikovsky, his other books, and his ideas for years.

This author has not read many books that were such well-documented, scholarly books, as Velikovsky's. And he wrote several of them — all in the days before personal computers and word processors. Amazing! He did an incredible amount of work to research and write those books. Yet he and his ideas were ridiculed, as they attempted to squelch his ideas.

We have freedom of speech here in the US, but when the intelligentsia or the elite decide they don't like what you have to say, they can make it very difficult to publish. We Americans may have these freedoms, but some still believe certain ideas deserve to be censored.

The idea today that the global warming debate is over because science has achieved *consensus*, is a form of censorship. Even in our free society, certain members of the elite believe it is too dangerous to allow this subject to be openly debated. If they can't stop publication, they may declare the debate "finished," which they hope will end its consideration as a news-worthy item. And when that happens, open discussions end.

Trouble is, this never seems to happen to obvious propaganda. Good ideas that should be discussed openly are squelched. Propaganda is printed and distributed freely.

Do we have choices here? We can support authors who are truthful. Their opinions should be openly discussed. Don't let anyone tell you whom or what you can and can't read. Take advantage of your freedoms while you still have them!

Take My Word For It ...

Finally, we are expected to take the words of many popular people, politicians, former politicians, etc. One or two of these people might actually be minimally qualified to voice their opinions. Most are not qualified. Does that stop them from speaking out? No!

For example, we are supposed to take the word of former VP Al Gore. He is riding on the laurels of his book, movie, academy award, and Nobel Peace Prize. The news media prefer his words over scientists who actually have some idea what they are talking about.

We are supposed to put absolute faith in the words of people like him, and anyone else who is popular. We are supposed to trust the news media. But we don't dare! Do not listen to anyone who answers skeptical questions like "Did global warming cause last week's heat wave? ... or last week's snow storm? ... or last week's *fill in the blank* ?" with answers that begin, "Of course!!"

Remember — in the year following Katrina --- we were supposed to experience several more even nastier hurricanes due to global warming? Nothing major happened that year. Following that quiet year, those who made the predictions were silent. But even so, they have continued predicting dire consequences from global warming. Eventually, even baseless predictions when repeated often enough will come true.

But now we learn from the *climategate* expert (the same guy whose research unit claimed that global warming was going to cause disastrous problems) that there has been no statistically significant warming in the past 15 years.[17]

The author could go on and on, but the point is clear. Spin is the name of the game. Everything is spun to fit the politically correct (PC) point of view. Anything that is not PC is treated as a joke — or disallowed, or silenced.

Our choice? Continue to exercise our rights as Americans. Don't let the alarmists tell you what you can and can't do. As an American, you have the right to decide for yourself. Expect that right! Do so!

---

[17] Petre, *loc.cit.*

Global Climate Change & God

God has been removed from so many topics in today's society that He is almost completely irrelevant to discussions — especially the discussion of climate change. Most people don't consider Him to play an active role in today's society. So He is hardly ever given a second thought — or even a first thought for that matter. He is just a figment of the imagination of a bunch of crazies. **But what if everything in the Bible is actually true?** What if we don't ignore Him? What does He have to say about these current issues?

Has God ever weighed in on these subjects? He had Moses write the story of creation in Genesis 1 & 2. It is unconstitutional to teach that in our schools. What else has God said that applies to climate change? Has anyone asked? Has anyone even considered what the Bible says about this subject?

The Bible is the most highly published and widely distributed book of all time. Almost everyone owns a Bible. Does anyone ever consult it on topics like this? **What does the Bible say about climate?** Who (if any) in the scientific community has asked this question? Generally speaking, the scientific community stays away from discussions of God.

The author has seen only a few such discussions, but not from the scientific community. Do the majority of Americans care what God thinks or says? We should! But that puts all who truly care into the realm of religion, faith, and miracles, and that, we are told, is kooky, dangerous, taboo, and certainly not PC.

The Bible is hardly ever mentioned in the context of these subjects which are all frequent topics of discussion in the mass media. According to Gonzalez and Richards[18], "The 'official' view now among scientists and academics is that the notion of **intelligent design** is either unscientific or at least superfluous to the practice of all natural science." (Emphasis added.) And that is just **intelligent design**. Those who support ID don't even offer a source for the intelligence. They simply want to **prove** (logically, of course) that intelligence was required in the creation of the universe.

Think what society would say to learn that **God is** the intelligent designer!! Actually, society seems to realize that God is the source — why else

---

[18] Gonzalez, Guillermo, and Richards, Jay W., The Privileged Planet, 2004, Regnery Publishing, Inc., Washington, DC, p.295.

would it be a taboo suggestion? If they mention God, ID supporters would not be allowed to participate in the public discussion. Even the slightest little hint that God had anything to do with creation would produce the same result. Opponents would shout, "That's religion!"

The suggestion that aliens supplied the intelligence is less controversial than the suggestion that God did. (We live in a crazy world!)

Many scientists who are Christians want to tiptoe through the PC minefields as they attempt to publish their findings — especially when their conclusions are consistent with creationism and the Bible. Many a career has been ruined[19] when a scientist has taken a stand contrary to accepted PC points-of-view. Their conclusions might be perfectly correct — **even scientifically accurate** — but political correctness rules in 21st Century America! And political correctness demands (consistent with 21st Century mores, I guess) that Christian men and women must not be allowed to speak.

This Book's Purpose

It is time for a scientist to ask the questions and consider what God has said about the earth, the world, and global climate change. This study will necessarily ask questions regarding creation, science, engineering principles, etc. As an engineer who is a Christian, the author not only sees the elegant designs of God everywhere in this creation, he sees many God-derived scientific principles at work every day.

In this book, we will consider some simple points of science that are related to the issue of climate. If God actually created the heavens and the earth, 21st Century science should find consistency between the evidence in the world around us and the words of the Bible that describe the earth and its workings. The two need not be mutually exclusive or totally contradictory as many would have us believe.

The first chapters in this book will consider God's words in the Bible that are applicable to these subjects. They will be followed by considerations of some relevant scientific principles, techniques, and phenomena. Our goal will be to evaluate these phenomena while explaining their roles in this debate and their relevance to the subject of global climate change.

---

[19] Stein, Ben, Expelled: No Intelligence Allowed, DVD, 21 Oct 08.

Ultimately, we want to decide if our children and grandchildren will really be in dire peril from man-made global warming? ... or if they will be in dire peril from mankind's arrogance?

Let us proceed.

2

Creation of Heaven & Earth

The first question to be addressed deals with creation. **According to the Bible, who created the heavens and the earth?**

The Bible is very clear on this point: **God created the heavens and the earth.** But skeptics require **scientific** proof. According to them, the Bible's explanation is a religious answer — it is fiction that requires faith — and faith is not science. Faith is simply not acceptable! But all explanations of creation require faith. The question is: In Whom do we place our faith? God? An unnamed intelligence source? Randomness? Chance?

The overwhelming popularity of evolution and almost total support by society do not make it correct. The fact that creationism is a religious belief that cannot be explained scientifically does not make it wrong. It just means that mankind puts more faith in logical, man-made arguments (logical to the minds of men) than to any words directly from the Creator God.

What exactly does the Bible say? As we look at the verses that answer this question, note that the statements that "**God created**" appear **throughout the whole Bible**. Genesis 1 and 2 focus on the creation, but the creation of the heavens and of the earth is addressed quite often throughout the rest of the scriptures.

Genesis 1 & 2

The first verse in the Bible answers the question directly: "**In the beginning God created the heaven and the earth.**"[20] The remainder of Genesis 1 gives details of the creation week. Genesis 2 describes the creation again with an emphasis on the creation of man and woman.

Why did God create man? Details of this answer appear in both chapters. Man was made in the "**image**" and "**likeness**" of God[21] to be God's representative on earth. God told man to, "**Be fruitful, and multiply, and <u>replenish</u> the earth, and <u>subdue</u> it: and <u>have dominion</u> over the fish of the sea, and over the fowl of the air, and over every living thing that moveth upon the earth.**"[22]

Note that Adam was told to fill the earth ("**replenish**" it) and take control of it ("**subdue**" it and "**have dominion**" over its inhabitants — the fish, the fowl, and the animals.) Genesis 2:15 reinforces this when God told Adam to "**dress**" and to "**keep**" the garden. That is, take care of it and protect it.

How were the universe, the earth, and all its inhabitants created? Gen 1.1 simply says that "**God created.**" God spoke — and it happened. "**<u>By the word of the LORD were the heavens made</u>; and all the host of them by the breath of His mouth.**"[23] "**Through faith we understand <u>that the worlds were framed by the word of God</u>, so that things which are seen were not made of things which do appear.**"[24] "**Let them praise the name of the LORD: for <u>He commanded, and they were created.</u>**"[25]

It is that simple. God spoke, and the worlds were created. Can we explain it in detail? No. Must we be able to? We can't. I trust that God

---

[20] Genesis 1:1

[21] Genesis 1:26-27

[22] Genesis 1:28

[23] Psalms 33:6

[24] Hebrews 11:3

[25] Psalms 148:5

created the worlds just as He said. Maybe someday, He'll explain it to me in more detail. Until then, I don't need more. You shouldn't either.

The Rest of the Old Testament

Although many people may think that the question "Who created?," is only answered in detail in the first two chapters of Genesis, many other Old Testament books explain that God created. Because mankind in general, and the Jews of the Old Testament, in particular, are forgetful, the words throughout the scriptures remind us repeatedly that God created the heavens, the earth, all creatures, all vegetation, and finally (and especially), mankind.

Note one other point. A variety of men wrote these books. We have first-hand reporting from many prophets of God whose jobs it was to accurately report God's words to mankind. God spoke to them, and they wrote it down for us. The multiple sources give much more credibility to the answer than just Moses' writings. To disbelieve the written record is to disbelieve the first-hand reporting of many prophets of God.

More from Moses

God created Adam in the garden in Genesis 1 & 2. Moses repeated this in Chapter 5 when he described Adam's whole family. "**This is the book of the generations of Adam. In the day that God created man, in the likeness of God made He him. Male and female created He them; and blessed them, and called their name Adam, in the day when they were created.**"[26]

Several years later when sin had overrun the whole creation, God decided to start again. He planned to save one man (Noah) and his family through the great flood. But even on that occasion, Moses reminded us with God's own words that He is the Creator. "**And the LORD said, 'I will destroy man whom I have created from the face of the earth; both man, and beast,**

[26] Genesis 5:1-2

and the creeping thing, and the fowls of the air; for it repenteth Me that I have made them.'"[27]

Nehemiah

Nehemiah knew very well who God was — and he recognized God as the Creator! "**Thou, even Thou, art LORD alone; Thou hast made heaven, the heaven of heavens, with all their host, the earth, and all things that are therein, the seas, and all that is therein, and Thou preservest them all; and the host of heaven worshippeth Thee.**"[28]

Nehemiah recorded in this communication that God made the heavens, the earth, the seas, and all that is therein. He also knew and made it clear to the Jews that God holds His creation together, "**Thou preservest them all.**" Not only that, but God receives worship from the "**host of heaven.**" All heavenly beings worship Him as Creator.

Job

"**There was a man in the land of Uz, whose name was Job; and that man was perfect and upright, and one that feared God, and eschewed evil.**" "**And the LORD said unto Satan, 'Hast thou considered My servant Job, that there is none like him in the earth, a perfect and an upright man, one that feareth God, and escheweth evil?'**"[29]

God said wonderful things about Job. But as we go through the book, we learn that Job thought he knew all about God — more, in fact, than he actually did. For all these good qualities, his attitude bordered on arrogance. In his time of great trial, he was confronted by God, who asked, "**Where wast thou when I laid the foundations of the earth? Declare, if thou hast understanding. Who hath laid the measures thereof, if thou knowest? Or**

---

[27] Genesis 6:7

[28] Nehemiah 9:6

[29] Job 1:1,8

who hath stretched the line upon it? Whereupon are the foundations thereof fastened? Or who laid the corner stone thereof."[30]

In this confrontation, God asked Job to explain all of the details of God's act of creating. To his friends, Job spoke like he was there, watching God as He created. But not so. He thought he knew much more about God than he actually did. So God asked him (rather indignantly, I suppose), "Where were you when I laid the foundations of the earth"??!!! God asked him an unbelievable series of questions regarding how He (God) created all life in the world. Job could say nothing. He wasn't there. He simply did not know.

It is doubtful you would hear any confrontation like that today because most people don't believe in the God of creation. Job knew God was the Creator, but he did not know any of the details of creation. We are in that same boat. We may be upright servants who believe that God is the Creator, but we don't have any idea how He did it, either. How could He possibly accomplish all that He did? Like I said above, maybe someday we will learn.

This is not to say that men today are not arrogant. They are! But unlike Job, in their arrogance today, many conclude that **God did not create — evolution, time, and random chance did**! That opinion certainly is insulting to the God of creation! So if, someday, you find yourself being questioned by God, it is best to take a pointer from Job — plead ignorance — say nothing.

Following his confrontation by God, Job was more careful about his statements and his thoughts. God's questions show very clearly that He formed the earth. He designed it. He measured it. He fastened its foundations. He laid its cornerstone. Job didn't know any of those details so he could not answer any of God's questions. The same applies to mankind today. We wouldn't be able to answer any of those questions either.

The Psalms

In the very next book of the Old Testament, the Psalmist inquired of the Creator's mindset when he asked God what was special about mankind? "**When I consider Thy heavens, the work of Thy fingers, the moon and the stars, which Thou hast ordained; What is man, that Thou art mindful of**

[30] Job 38:4-6

him? and the son of man, that Thou visitest him? For Thou hast made him a little lower than the angels, and hast crowned him with glory and honour. Thou madest him to have dominion over the works of Thy hands; Thou hast put all things under his feet: All sheep and oxen, yea, and the beasts of the field; the fowl of the air, and the fish of the sea, and whatsoever passeth through the paths of the seas."[31]

Note all the details of the Psalmist's knowledge evidenced in his questions. The Psalmist knew that God not only created the heavens, the earth, the moon, the stars, and all living creatures, but he knew that God created man in a very special way. The Psalmist knew that man was given dominion over every living being in the creation — over all living creatures on earth, in the air, and in the sea. The Psalmist didn't know why or how God did this, but he knew the fact that God did it. And he knew that mankind had been given special responsibilities.

In a later chapter, the Psalmist informs us that God **spoke** — and the creation came into being. **"By the word of the LORD were the heavens made; and all the host of them by the breath of His mouth. He gathereth the waters of the sea together as an heap: He layeth up the depth in storehouses. Let all the earth fear the LORD: let all the inhabitants of the world stand in awe of Him. For He spake, and it was done; He commanded, and it stood fast."**[32]

By God's "**word**" and "**the breath of His mouth**," God made the heavens. And He gathered together the waters of the sea. Anyone who has stood at the seashore or ridden on a ship or flown over an ocean in an airplane has experienced the sheer enormity of earth's seas. The author has flown around the world several times and seen firsthand that the oceans are absolutely enormous. The idea that God could gather up the waters, as one would gather up a handful of sand on the beach, is mind-boggling. Yet the Psalmist tells us He did! And mankind's response should be to "**fear the Lord**," and to "**stand in awe of Him**." All God needed to do was speak — and it was done! Amazing!

---

[31] Psalms 8:3-8

[32] Psalms 33:6-9

We learn further from the Psalmist that God used His hands to form the heavens and the earth. **"The sea is His, and He made it: and His hands formed the dry land."**[33] **"Of old hast Thou laid the foundation of the earth: and the heavens are the work of Thy hands."**[34]

As large as the oceans are, the earth is even larger. Even in 747s, it takes almost four days to fly around the globe. In the space shuttle, it takes a little over an hour per orbit. Nevertheless the Psalmist refers to God's molding of the earth and the heavens in His hands. How could anyone not stand in awe of such a God? ... unless, of course, they refuse to believe in such a God.

The Psalmist very clearly attributed the whole creation to God. It is all His! **"The heavens are Thine, the earth also is Thine: as for the world and the fulness thereof, <u>Thou hast founded them</u>."**[35] God founded them.

The reminders continue in the midst of Psalms of praise to God "<u>**who laid the foundations of the earth, that it should not be removed for ever.**</u>"[36] **"To Him that by wisdom <u>made the heavens</u>: for His mercy endureth for ever. To Him <u>that stretched out the earth above the waters</u>: for His mercy endureth for ever. To Him that <u>made great lights</u>: for His mercy endureth for ever."**[37] **"... whose hope is in the LORD his God: <u>Which made heaven, and earth, the sea, and all that therein is</u>: which keepeth truth for ever:"**[38]

Later in the Psalms, which contain many songs of praise, the Psalmist continues to praise God for His creation. God laid the foundations of the earth; He made the heavens; He stretched out the dry land above the waters; He made the great lights in the sky; and in summary, God made everything — heaven, earth, the sea, and all life.

---

[33] Psalms 95:5

[34] Psalms 102:25

[35] Psalms 89:11

[36] Psalms 104:5

[37] Psalms 136:5-7

[38] Psalms 146:5-6

All of creation — the angels, the heavenly hosts, the sun, moon, stars, heavens, and the water above the heavens — are all called upon to praise the Lord because He created. **"Let them praise the name of the LORD: <u>for He commanded, and they were created</u>."**[39]

Proverbs

In the Proverbs, Solomon explains that the creation is not mindless, random, or evolving. God created with **<u>wisdom</u>**. Wisdom speaks (wisdom is the "I") in this passage from the 8th chapter of Proverbs: **"I was set up from everlasting, from the beginning, or ever the earth was. When there were no depths, I was brought forth; when there were no fountains abounding with water. Before the mountains were settled, before the hills was I brought forth: While as yet He had not made the earth, nor the fields, nor the highest part of the dust of the world. <u>When He prepared the heavens, I was there: when He set a compass upon the face of the depth: When He established the clouds above: when He strengthened the fountains of the deep: When He gave to the sea his decree, that the waters should not pass His commandment: when He appointed the foundations of the earth</u>: Then I was by Him, as one brought up with Him: and I was daily His delight, rejoicing always before Him; Rejoicing in the habitable part of His earth; and my delights were with the sons of men."**[40]

From this passage, we learn that God used His wisdom to design and create the heavens, the earth, and the universe. This is very different from that which the proponents of evolution would have us believe. They want us to believe the creation occurred totally randomly, without the need for any intelligence — by chance. After looking only at the incredibly complex and elegant design of the human body, it is inconceivable that such a design happened randomly or by sheer chance. Intelligence and wisdom are evident in the human design — and in the design of all other life as well. It is sheer arrogance on man's part to deny any intelligence whatsoever in the formation of life on this planet. But that doesn't stop many from doing so.

---

[39] Psalms 148:5

[40] Proverbs 8:23-31

Isaiah

More reminders come from the pen of the prophet Isaiah, who also attributed the creation to God. **"Thus saith God the LORD, He that created the heavens, and stretched them out; He that spread forth the earth, and that which cometh out of it; He that giveth breath unto the people upon it, and spirit to them that walk therein."**[41] **"Hast thou not known? Hast thou not heard, that the everlasting God, the LORD, the Creator of the ends of the earth, fainteth not, neither is weary? There is no searching of His understanding."**[42]

Here again, Isaiah attributed the creation of the heavens, the earth, and all living beings to God. Specifically, he gives credit for our breath and spirit — our lives — to God. In the second passage, he highlights the power of God, who doesn't faint, neither is weary, in the great work of creation. Isaiah also testified to the fact that God knew what He was doing when He created — all of His understanding went into the design.

The task of any prophet of God is to accurately relay God's words to mankind. In Isaiah's case, he quoted God directly – using God's own words to remind everyone who He is and what He has done. God said, **"I form the light, and create darkness: I make peace, and create evil: I the LORD do all these things. Drop down, ye heavens, from above, and let the skies pour down righteousness: let the earth open, and let them bring forth salvation, and let righteousness spring up together; I the LORD have created it."**[43]

This passage goes a little further with information about the creation. Isaiah quoted God's exact words, when He informed the people that He created light, darkness, peace, and evil. God's goal, stated in this passage, is to see righteousness and salvation dwell abundantly on the earth that He created.

The 45th Chapter of Isaiah gives us another important detail of the creation. God formed the earth **to be inhabited**. He designed it specifically to be under the dominion of mankind. It was not an accident that He made it the

---

[41] Isaiah 42:5

[42] Isaiah 40:28

[43] Isaiah 45:7-8

way He did. He is God; He decided it should be this way; and He formed it accordingly.

"**I have made the earth, and created man upon it: I, even my hands, have stretched out the heavens, and all their host have I commanded.**"[44] "**For thus saith the LORD that created the heavens; God Himself that formed the earth and made it; He hath established it, He created it not in vain, He formed it to be inhabited: I am the LORD; and there is none else.**"[45]

Lest anyone doubt, it was God who "**laid the foundation of the earth.**" It was God's "**right hand**" who "**spanned the heavens.**" And it is God Himself who commands and controls His creation. When He calls, the whole creation pays attention.

"**Mine hand also hath laid the foundation of the earth, and My right hand hath spanned the heavens: when I call unto them, they stand up together.**"[46]

In these quotations of God's own words, note all of the personal pronouns. God said, "**I** created!"; "**I** made the earth and man upon it"; "**I** stretched out the heavens"; "**My** hand laid the foundation of the earth"; and "**My** right hand spanned the heavens." "**I** am the LORD; and there is none else." God created the earth — not in vain — but to be inhabited! He put lots of wise decisions into His design of the earth, which He created for mankind.

When the people forgot God (which happened quite often then, as it happens quite often today), Isaiah directly confronted them. "**And forgettest the LORD thy maker, that hath stretched forth the heavens, and laid the foundations of the earth; and hast feared continually every day because of the fury of the oppressor, as if he were ready to destroy? and where is the fury of the oppressor?**"[47]

---

[44] Isaiah 45:12

[45] Isaiah 45:18

[46] Isaiah 48:13

[47] Isaiah 51:13

Why should the people then be paying attention? Why should we today be paying attention? Because the Creator, the Lord, is our maker. He created the heavens and the earth and He watches out for our well being.

Isaiah went even further. The Creator God wanted the Jews then and Christians today to be His people. He created the heavens and the earth for His people. And not only for His people alone — He wants all of mankind to be His bride. And He will be their husband. This signifies that He wants a close familial relationship with mankind. Who shall be in this relationship with the Creator? "**The whole earth.**" His plan included everyone!

"**And I have put My words in thy mouth, and I have covered thee in the shadow of Mine hand, that I may plant the heavens, and lay the foundations of the earth, and say unto Zion, 'Thou art My people.'**"[48] "**For thy Maker is thine husband; the LORD of hosts is His name; and thy Redeemer the Holy One of Israel; The God of the whole earth shall He be called.**"[49]

As God's prophet, Isaiah bluntly reminded the people that God not only created — He is the "**God of the whole earth**"! "Remember! Don't forget!" To whom is/was Isaiah talking? In his day, he was talking directly to the people of Israel. Today, his words apply to the rest of us. Not everyone will pay any attention, but many surely will. God created this wonderful world, planet, and all life on it for the benefit of mankind! Isn't that great?!

Jeremiah

Jeremiah's prophecy clarified that mankind's idols did **not** create the heavens or the earth. According to Jeremiah, God used His power and wisdom to create the earth and the world, and the heavens were stretched out at His discretion: "**Thus shall ye say unto them, 'The gods that have not made the heavens and the earth, even they shall perish from the earth, and from under these heavens. He hath made the earth by His power, He hath established the world by His wisdom, and hath stretched out the heavens by His discretion. When He uttereth His voice, there is a multitude of**

---

[48] Isaiah 51:16

[49] Isaiah 54:5

waters in the heavens, and He causeth the vapours to ascend from the ends of the earth; He maketh lightnings with rain, and bringeth forth the wind out of His treasures.'"[50]

Throughout the Bible, we find various peoples worshiping many different gods. None of those idols (their gods) were capable of creating. Those idols were simply inanimate objects of wood, metal, or stone. But the Creator God used His wisdom and understanding to establish the world.

Tying wisdom and understanding (discretion) to the act of creation is very different than what one hears from evolutionists who believe we came into being due to randomness and chance. I, for one, prefer to believe that mankind is a product of God's wisdom and understanding, consistent with these words from Jeremiah. If you want to believe that mankind is a product of randomness and chance, that is your decision to make.

Amos

One of the minor prophets, Amos, declared that "**The Lord, the God of hosts**" formed the mountains and created the wind. "**For, lo, <u>He that formeth the mountains, and createth the wind</u>, and declareth unto man what is His thought, <u>that maketh the morning darkness, and treadeth upon the high places of the earth</u>, The LORD, The God of hosts, is His name.**"[51]

Following this, Amos added that "**The Lord**" made the stars and constellations and poured the waters of the sea upon the earth. "**<u>Seek Him that maketh the seven stars and Orion, and turneth the shadow of death into the morning, and maketh the day dark with night: that calleth for the waters of the sea, and poureth them out upon the face of the earth</u>: The LORD is His name:**"[52]

Amos continued: God placed His people on earth under the canopy of heaven and poured the sea's waters onto the earth. "**It is He that buildeth His stories in the heaven, and hath founded His troop in the earth; He that**

---

[50] Jeremiah 10:11-13

[51] Amos 4:13

[52] Amos 5:8

calleth for the waters of the sea, and poureth them out upon the face of the earth: The LORD is His name."[53]

Zechariah

In Zechariah's prophecy, he also declared that "**the Lord**" is responsible for creating the heavens and the earth, and for placing the spirit of man within mankind. **"'The burden of the word of the LORD for Israel,' saith the LORD, 'which stretcheth forth the heavens, and layeth the foundation of the earth, and formeth the spirit of man within him.'"**[54]

Very clearly in many places throughout the Old Testament, the writers, who were all prophets of God, claimed that the Lord God was responsible for the creation of the heavens, the earth, and all life that resides therein. If Moses alone had made such a declaration, one might wonder if it was true. But we have just quoted from eight other Old Testament books, the authors of which all declare the Lord God to be the Creator. The evidence doesn't stop there, though. It continues in the New Testament.

The New Testament

Throughout the Bible, over and over again, mankind is told that God created the heavens and the earth. No one else in the universe had the wisdom, understanding, power, or might to create the heavens and the earth. God did it! Neither were the worlds created by chance — God used His omniscience to create all things.

This message does not stop when we pass from the writings of the Old Testament to the New Testament. It continues to be repeated.

The Gospel of John

John, in his gospel, made it clear who created the worlds. **"In the beginning was the Word, and the Word was with God, and the Word was**

---

[53] Amos 9:6

[54] Zechariah 12:1

God. The same was in the beginning with God. All things were made by Him; and without Him was not any thing made that was made."[55]

The Word, whom we learn a few verses later (in John 1:14) is Jesus, was the Creator. He is eternal in His being; He was with God in the beginning; He created all things; and nothing was created apart from Him.

The Acts of the Apostles

Luke, too, declared God to be the Creator and Lord of heaven and earth. "**God that made the world and all things therein, seeing that He is Lord of heaven and earth, dwelleth not in temples made with hands;**"[56] "**Lord of heaven and earth**" refers to God's position as the One who deserves all worship and praise, and the One who controls all.

Ephesians

The God of the Bible is a triune God (3 parts in one): Father, Son, and Holy Spirit. In the New Testament, we learn that the Son, Jesus Christ, is the Creator of the heavens and the earth. "**And to make all men see what is the fellowship of the mystery, which from the beginning of the world hath been hid in God, who created all things by Jesus Christ.**"[57]

After Jesus appeared on the scene — on earth, that is — we learn that Jesus Christ was the Creator. A lot of details like this come from the pen of Paul. Initially, Paul's given name was Saul.

Saul was a Roman citizen, educated in Judaism as a Pharisee — one of the religious elite of that day. He spent most of his time during the early days of his career chasing and persecuting Christians. But then, one day on the Damascus road, he was confronted by Jesus. As a result, he became a follower of Christ; his name was changed to Paul; and he spent the rest of his life spreading God's words among the people.

---

[55] John 1:1-3

[56] Acts 17:24

[57] Ephesians 3:9

The name Saul means *asked for*; the name Paul means *little*. Only after Paul became little in his own eyes did he become of great use to God. Everywhere he went, he preached first to the Jews but then also to the Gentiles. His letters form much of the New Testament. We can learn a lot about God and His Son, Jesus Christ, by reading the letters of Paul.

Colossians

Speaking about the Son in his letter to the Colossians, Paul declared, "**For by Him were all things created, that are in heaven, and that are in earth, visible and invisible, whether they be thrones, or dominions, or principalities, or powers: all things were created by Him, and for Him. And He is before all things, and by Him all things consist.**"[58]

This passage takes us further than any of the passages in the Old Testament. According to Paul, this world in which we live is God's world. He created it! He maintains it! He is preeminent! But Paul takes it one step further. All things in heaven and earth were created by God and for God. The list includes earthly things, as well as heavenly things — heavenly beings — thrones, dominions, principalities, and powers. This includes all heavenly beings such as angels, fallen angels, and powers that be — of which we humans know little.

Not only does Paul say that God created all things, but he says that God stands before all things (that is, He rules all) and He holds all together as we go throughout this life. The world cannot fly apart at any point because God is holding it together — "**by Him all things consist.**"

This is rather incredible! The very God of creation is holding the whole universe together so that all the skeptics and naysayers of any kind can express their opinions and have this discussion concerning who created! The one person, whom many deny had anything to do with the creation, is maintaining the integrity of the universe to make it possible for skeptics to have this argument and voice their opinions — that is, to deny He exists.

---

[58] Colossians 1:16-17

The Letter to the Hebrews

The writer of the letter to the Hebrews explains that God "**hath in these last days spoken unto us by His Son, whom He hath appointed heir of all things, by whom also He made the worlds.**"[59] "**And, Thou, Lord, in the beginning hast laid the foundation of the earth; and the heavens are the works of thine hands.**"[60] "**Through faith we understand that the worlds were framed by the word of God, so that things which are seen were not made of things which do appear.**"[61]

Here again, when writing to the Hebrew people, the writer makes it clear whom the Creator really is. It is the Son of God who was appointed heir of all things and who created all things. It is the Messiah, whom the Jews are still waiting to identify. Unfortunately, they refuse to read the New Testament. They even ignore letters like this one to the Hebrews, which was meant specifically for them and which answers many of their questions.

The other two verses repeat what we heard over and over again in the Old Testament. God created the heaven and the earth. And a little more specifically, the Word of God created the heavens and the earth. God spoke and the heavens and the earth were created.

The Revelation of John

The Apostle John repeated these same ideas: God created all things; and He created them for His own pleasure. "**Thou art worthy, O Lord, to receive glory and honour and power: for Thou hast created all things, and for Thy pleasure they are and were created.**"[62]

We understand that John was not only the youngest of the Apostles, but he was very close to Jesus. In his old age, he wrote the Book of Revelation where he makes clear that his beloved friend was the Creator. And he adds

---

[59] Hebrews 1:2

[60] Hebrews 1:10

[61] Hebrews 11:3

[62] Revelation 4:11

that the creation occurred for Jesus' pleasure. What pleasure was that? The triune God wanted mankind to be His people — His companions — His family. Not just a few men and women here and there — He wanted them all to enjoy this status with Him! And so He created this wonderful world in which mankind could live and have dominion.

In a later chapter in the Revelation, John described an angel who lifted up his hand to heaven, "**And sware by Him that liveth for ever and ever, who created heaven, and the things that therein are, and the earth, and the things that therein are, and the sea, and the things which are therein, that there should be time no longer.**"[63] That angel was a messenger sent by God, the Creator.

By the 10th Chapter of the Revelation, judgment was about to be brought down upon all those men and women who rejected God. But even then, the angel reminded the people that what was about to occur was coming from the Creator.

Summary

Consider all of the verses referenced in this chapter. They contain lots of direct, first-hand evidence from more than a dozen of God's servants, several in God's own words, that clearly indicate that God created the world and all that is in it. The Bible is a history book written by prophets of God. Its writers attribute the creation to God. It doesn't just say it once in a single obscure verse, or only in the two creation chapters in Genesis. It repeats it over and over again throughout the whole canon of scripture.

These are not eye-witness accounts of the creation because none of the writers were present when the worlds were created. As Prophets of God, however, many of these verses represent direct testimony from God to prophet to mankind. Remember: the mark of a true prophet of God is 100% accuracy in the transmission of God's words to mankind. For this reason alone, we should be paying close attention to their words.

Who created the heavens and the earth? According to the Bible, God did. It wasn't chance. It wasn't randomness. It wasn't evolution. It was God who created. It can't get any simpler and more clear than that.

---

[63] Revelation 10:6

3

Control of Heaven & Earth

The second question that needs to be asked is: **According to the Bible, who is now in control of the heavens and the earth?** The simple answer, which we have already seen in the previous chapter, is: **The same God who created the heavens and the earth controls it.** Where and how does the Bible explain this?

Direct Evidence

The Whole Creation is God's

Direct support for this answer comes from verses such as these two, written by the Psalmist: "**The earth is the LORD'S, and the fulness thereof; the world, and they that dwell therein.**"[64] "**Whatsoever the LORD pleased, that did He in heaven, and in earth, in the seas, and all deep places.**"[65]

The writer of the first book of Chronicles also proclaimed this. "**Thine, O LORD, is the greatness, and the power, and the glory, and the victory, and the majesty: for all that is in the heaven and in the earth is Thine; Thine is the kingdom, O LORD, and Thou art exalted as head above all.**"[66]

---

[64] Psalms 24:1

[65] Psalms 135:6

[66] 1 Chronicles 29:11

The earth in all its fullness — that is, everything in, of, and about the earth — is the Lord's. Not only is everything His, but He does whatsoever He pleases in it. In other words, He controls everything. He can allow His creation to proceed without His intervention, or He can intervene as He wishes. That's what it means to be in control.

God Commands the Host of Heaven

As Creator, God commands the host of heaven: **"I have made the earth, and created man upon it: I, even My hands, have stretched out the heavens, and all their host have I commanded."**[67]

This takes God's control beyond mere control of the earth. It takes His control into the heavens where He commands the host of heaven. This could refer to heavenly beings and powers, and it could refer to the stars in the sky. Either way, God's command is powerful and far-reaching.

God Holds All Together

Paul wrote concerning God's "**dear Son,**" **"For by Him were all things created, that are in heaven, and that are in earth, visible and invisible, whether they be thrones, or dominions, or principalities, or powers: all things were created by Him, and for Him: And He is before all things, and by Him all things consist."**[68] God's Son, Jesus, created all things and "**by Him all things consist.**" That is, He holds everything together and keeps the world functioning properly by His power.

God's Control Extends Throughout the Universe

God's power extends far beyond what man imagines. When Jesus stood at trial before Pilate, **"Then saith Pilate unto Him, 'Speakest Thou not unto me? Knowest Thou not that I have power to crucify Thee, and have power to release Thee?' Jesus answered, 'Thou couldest have no power at**

---

[67] Isaiah 45:12

[68] Colossians 1:16-17

all against Me, except it were given thee from above: therefore he that delivered Me unto thee hath the greater sin.'"[69]

Pilate thought he was in charge, but Jesus told him that was an incorrect assumption. The only reason Pilate had any power at all was that it had been given to him from above. Who was in charge? Who was in control? God was.

God is more powerful in daily happenings on earth than anyone imagines. God is more powerful throughout the universe than anyone imagines. It is the Creator, who controls His creation.

God Preserves

God preserves all of His created beings and the host of heaven worship Him for it. Without God's help, mankind would have major problems just existing. But God is preserving life as we know it. **"Thou, even Thou, art LORD alone; Thou hast made heaven, the heaven of heavens, with all their host, the earth, and all things that are therein, the seas, and all that is therein, and Thou preservest them all; and the host of heaven worshippeth Thee."**[70]

The Soul of Every Living Thing Is In God's Hand

Zophar the Naamathite asked Job about his fallen state. **"Canst thou by searching find out God?"**[71] This is a small part of a long drawn out question and commentary by Zophar. Job's answer was equally long and convoluted. **"But ask now the beasts, and they shall teach thee; and the fowls of the air, and they shall tell thee: Or speak to the earth, and it shall teach thee: and the fishes of the sea shall declare unto thee. Who knoweth**

---

[69] John 19:10-11

[70] Nehemiah 9:6

[71] Job 11:7a

not in all these that the hand of the LORD hath wrought this? In whose hand is the soul of every living thing, and the breath of all mankind."[72]

Job declared that the whole creation knows — that the beasts, the fowls, the fishes of the sea, and even the earth itself understand. Each understands well enough to teach it. What do they understand? They know that the hand of the Lord hath wrought this, and "**That the soul of every living thing, and the breath of all mankind**" are in God's hand. God is in control! The whole creation knows.

Men, who supposedly are most intelligent, don't seem to know this. Men, who are spiritual beings, don't seem to know this. Men insist on using their own smarts to understand and explain everything — and they simply don't understand.

There is an expression — "dumb as a box of rocks." Well, as part of the earth, in this case even the rocks know God created and God controls.

It should be very clear. Our life, and every breath we take, is in the hand of God. He is in control.

God Can Control Whomever He Wants

God told Cyrus, King of Persia, "**I am the LORD, and there is none else, there is no God beside Me: I girded thee, though thou hast not known Me: That they may know from the rising of the sun, and from the west, that there is none beside Me. I am the LORD, and there is none else. I form the light, and create darkness: I make peace, and create evil: I the LORD do all these things.**"[73]

Whatever God wants to do (or needs to do) on this earth is accomplished quickly. Notice when God talked to Cyrus, King of Persia, who was not one of His people, that He ended the verse with "**though thou hast not known Me.**" Cyrus was a foreign king, whom God was about to send against His own people, the children of Israel. God is in control, and He exercises that control on whomever He needs — even on foreign (non-Jewish) kings. God has total control over His whole creation — that includes everyone.

---

[72] Job 12:7-10

[73] Isaiah 45:5-7

God Controls — Mankind Has Dominion

Although God is in control, He gave man **dominion** over all of the living creatures that move upon the earth in Genesis 1:26. He told Adam to **replenish** the earth, to **subdue** it, to **have dominion** over it, to **dress** it, and to **keep** it.[74]

Because God placed man on the earth and gave him dominion over it, we read statements like the following: **"The heaven, even the heavens, are the LORD'S: but the earth hath He given to the children of men."**[75]

With this dominion by man in mind, the Psalmist asked God to explain His special interest in mankind. **"O LORD our Lord, how excellent is Thy name in all the earth! Who hast set Thy glory above the heavens. Out of the mouth of babes and sucklings hast Thou ordained strength because of Thine enemies, that Thou mightest still the enemy and the avenger. When I consider Thy heavens, the work of Thy fingers, the moon and the stars, which Thou hast ordained; What is man, that Thou art mindful of him? and the Son of Man, that thou visitest Him? For Thou hast made him a little lower than the angels, and hast crowned him with glory and honour. Thou madest him to have dominion over the works of Thy hands; Thou hast put all things under his feet: All sheep and oxen, yea, and the beasts of the field; The fowl of the air, and the fish of the sea, and whatsoever passeth through the paths of the seas. O LORD our Lord, how excellent is Thy name in all the earth!"**[76]

A full explanation of this passage goes well beyond the scope of this study. Suffice it to say that man was supposed to take control over the earth as God's representative — i.e., to have **dominion.** But it didn't happen. Adam didn't accomplish this. Adam sinned, mankind fell, Adam and Eve were kicked out of the Garden of Eden, and God cursed the earth for Adam's sake.[77]

---

[74] Genesis 1:28 and 2:15

[75] Psalms 115:16

[76] Psalms 8:1-9

[77] Genesis 3:17

From that point on, neither Adam nor his descendants were in a position to take dominion as originally planned.

The parallel passage to this Psalm 8 passage is in Hebrews 2. According to this passage, man (specifically, Adam) did not successfully take dominion. "**But one in a certain place testified, saying, 'What is man, that Thou art mindful of him? or the Son of Man, that Thou visitest him? Thou madest him a little lower than the angels; Thou crownedst him with glory and honour, and didst set him over the works of Thy hands: Thou hast put all things in subjection under his feet.' For in that He put all in subjection under him, He left nothing that is not put under him. *But now we see not yet all things put under him.* But we see Jesus, who was made a little lower than the angels for the suffering of death, crowned with glory and honour; that He by the grace of God should taste death for every man.**"[78]

Quoting from Psalm 8, it explains, "**But now we see not yet all things put under him.**" Adam did not take the dominion he was given by God. But all was not lost. God sent His Son, the man Jesus, as a new representative of God on earth. The man, Jesus, **took dominion** as God's representative on earth . He accomplished what Adam could not.

According to 1 Corinthians 15:45-47, Adam was "**the first man**" who was "**of the earth**"; Jesus was "**the second man**" who is "**the Lord from heaven.**" Adam, or course, was the first Adam. This passage describes Jesus as "**the last Adam.**" God was not going to send any more representatives to earth. God created Adam, who sinned and failed God's mandate. He then sent His Son, the man Jesus, who did fulfill mankind's mandate. The first Adam failed; the last Adam did not; no other Adam is needed.

The only representative of God, to date, who has actually taken dominion over the earth as planned, is Jesus. When Jesus returns to earth at His second advent, the rest of mankind (that is, the children of God) will finally take proper form and also fulfill this mandate. But until then, mankind has not and will not take complete dominion over the earth.

Mankind may think they have dominion, but they do not. Mankind has been able to accomplish many wonderful technological feats on this earth,

---

[78] Hebrews 2:6-9

but strictly speaking, fallen man does not have dominion. But God is in total control.

Circumstantial Evidence

Practically everything else in the Bible gives circumstantial support to the fact that God controls heaven and earth. This only needs brief mention here. A complete study of this point is left as homework to each reader.

Consider how God controlled every part of the creation and everything that transpired in the garden. He laid down the rules for Adam and Eve. They disobeyed, He kicked them out of the garden, and He set up guards to prevent their return. God was in charge in the garden.

Skipping forward to the time when the Jews were enslaved in Egypt, God brought them out. God's call to Pharaoh was, **"Let My people go."** He warned the Egyptians as He sent numerous plagues among them. All of the signs were to identify Him as God. He warned them again, before He passed over and killed all male firstborn in the land.[79] Finally, Pharaoh told Moses that the people could go. Then, God divided the sea so the Israelites could pass over on dry land, but He closed the sea upon the Egyptian armies when they tried to follow. God was totally in charge of all those events.

In the desert, God provided guidance, water, and food, to the wandering Israelites. God was in charge of their wandering in the desert.

God told Jeremiah, **"Before I formed thee in the belly I knew thee; and before thou camest forth out of the womb I sanctified thee, and I ordained thee a prophet unto the nations."**[80] God controlled Jeremiah's forming in the womb, just as He has similar control over the formation of everyone else who comes into this world.

Jesus came to earth according to God's plan: **"For unto us a child is born, unto us <u>a son is given</u>."**[81] Note that the **son** was given; and the **child** was born. Isaiah was very clear on this point. It required that the Father in

---

[79] Exodus 4:22-23

[80] Jeremiah 1:5

[81] Isaiah 9:6

heaven send His Son — and He did. Again, God demonstrated that He is in charge.

In the New Testament, we can look at any of the miracles that Jesus performed. He was in charge. The demons whom Jesus encountered in His many travels all knew who He was — they identified Him as the Son of God — and they obeyed Him.

Jesus healed many during His time of ministry on earth. Matthew told the story of one particular leper: **"And, behold, there came a leper and worshipped Him, saying, 'Lord, if Thou wilt, Thou canst make me clean.' And Jesus put forth His hand, and touched him, saying, 'I will; be thou clean.' And immediately his leprosy was cleansed."**[82] Only the God from heaven can do such things. He was in control of the situation. **"I will,"** He said. And the leprosy was cleansed.

Jesus Himself said frequently throughout John's Gospel that He was only acting according to God the Father's desires. Jesus healed the sick, raised the dead, turned loaves and fishes into feasts for the multitudes, turned water into wine, walked on water, calmed the winds and the seas, etc. God was in charge and He did as it pleased Him. He is in control.

When Jesus confronted Saul in Acts 9, Saul bowed down and obeyed. After Saul's conversion and his name-change to Paul, he wrote many letters to the churches and clearly taught that God was in charge of everything.

Summary

Throughout the whole Bible, histories are recorded by prophets of God, words of God are relayed, directions are given, and mankind is taught that the God of creation is in charge of the heavens, the earth, and all life in the world.

There are simply too many passages that cover this issue to go into in any great detail here. Whenever any information was needed throughout the whole Bible, men of God contacted Him and He told them what was about to happen, or what to do. The whole Bible stands as evidence that God is in control of the heavens, the earth, and every living creature in it. He holds all of mankind's lives in His hand. And as for all other living creatures — He cares for and maintains them, too.

---

[82] Matthew 8:2-3

Who is now in control of the heavens and the earth? The Bible clearly teaches us that God is in control of everything. Mankind may think we are in control, but that would be wrong. Americans may think we are in control, but that would be wrong. It is only the arrogance of man that leads to doubt and disbelief that God is actually in control.

Control of Wind & Rain

In Chapter 2, we showed that the Bible declares God to be the Creator of the heavens and the earth. Although the story of creation is known to be in Genesis 1 and 2, recognition that God created can be found over and over again throughout the whole Bible.

The question concerning who **controls** the heavens and the earth was studied in Chapter 3. Once again, the Bible shows that the heavens and the earth are under God's control. He does whatever He wants with anything and everything in the creation.

But where do we see most of God's control throughout the Bible? And what parts of His control impact the subjects of global warming and climate change? We see those issues in His control over the wind and the rain — which together form the weather and the climate.

So in this chapter, we ask the question: **According to the Bible, who controls the wind and the rain?** (... as well as the hail, the snow, the storms, the lightning, etc.?) The Bible indicates, as you will see, that **God controls the wind and the rain.**

Many say the Bible is a "non-scientific" document. Its understanding requires faith and faith is not science. So therefore, society would have you to believe that anything the Bible says is irrelevant to the global warming discussion. There is no conceivable value in asking what the Bible says — at least according to society, that is.

Many who claim the Bible is a non-scientific, irrelevant document are the same people who teach that evolution is a proven science. Rather than ignore the Bible, we need to consider **if** it says anything about the climate, and if so, **what** it actually says.

The Bible appears to be the last place anyone has looked for answers regarding climate change. Most people would think that the Bible has absolutely nothing to say about climate, but what if most people are wrong? What if mankind does not have a current lock on the understanding of climate systems because they have spent their whole lives using their own limited mental abilities to answer questions on that subject?

This chapter will focus on Bible passages that apply to the climate, in general, and the wind, rain, hail, lightning, and storms, in particular. **What does the Bible say? Who controls the wind and the rain?**

Early Conditions Following the Creation

There is no mention of rain, snow, hail, nor storms from the time Adam was created until the time of Noah and the flood in Genesis 6. In those days, humans and animals appear to have been vegans. There was obviously a water supply (**"a mist"**) over the earth as well as streams of water springing up from the earth. We are left to wonder what, if anything, else was present in those days.

"But there went up a mist from the earth, and watered the whole face of the ground."[83] **"And a river went out of Eden to water the garden; and from thence it was parted, and became into four heads."**[84]

Many think this mist was a great canopy of water vapor that enveloped the earth. Springs of water out of the earth formed rivers of water. And between the two (mists and ground waters), the people, plants, and animals had sufficient water to live and to grow.

The Great Flood

But then came the Great Flood — the Deluge. In Genesis 6:17, God told Noah to build an ark because He was going to bring **"a flood of waters upon the earth."** **"And, behold, I, even I, do bring a flood of waters upon**

---

[83] Genesis 2:6

[84] Genesis 2:10

the earth, to destroy all flesh, wherein is the breath of life, from under heaven; and every thing that is in the earth shall die."[85]

The first mention of "rain" in the Bible comes later in Genesis 7.4. **"For yet seven days, and I will cause it to rain upon the earth forty days and forty nights; and every living substance that I have made will I destroy from off the face of the earth. "**[86]

One wonders whether Noah understood the meanings of "**a flood of waters**" or "**rain**", but he obeyed God nevertheless. And it was very clear from God's words that **He** was going to do this.

After first hearing the word *flood*, Noah spent a century building the ark. It was a long time and he exhibited lots of faith. Imagine being asked over and over again, over a period of 100 years, "What are you building Noah? Why???"

After first hearing the word *rain*, however, Noah didn't have long to wait: "**And it came to pass after seven days, that the waters of the flood were upon the earth. In the six hundredth year of Noah's life, in the second month, the seventeenth day of the month, the same day were all the fountains of the great deep broken up, and the windows of heaven were opened. And the rain was upon the earth forty days and forty nights."**[87]

"**All the fountains of the great deep were broken up, and the windows of heaven were opened.**" On this occasion, God brought water onto the earth from every possible source — from above and from below.

Note that the flood was predicted and brought by God. God's words were, "**And, behold, I, even I, do bring a flood of waters upon the earth,**" and "**I will cause it to rain upon the earth forty days and forty nights.**" He clearly told Noah that He [God] was going to bring a flood of waters, and that He [God] was going to cause it to rain for 40 days and 40 nights. That is pretty good weather prediction and pretty good control over the wind and rain. The flood and the rain came, the storm happened as predicted, and its duration lasted exactly 40 days and 40 nights.

---

[85] Genesis 6:17

[86] Genesis 7:4

[87] Genesis 7:10-12

God said to Noah, "**I do bring a flood,**" and "**I will cause it to rain.**" This shows very clearly that God controlled the storm, the rain, and the fountains of the deep. He had no problem telling Noah what He was about to do. And it happened — exactly as He said it would.

After the flood, we learn of new capabilities of God and new conditions on earth. "**And God remembered Noah, and every living thing, and all the cattle that was with him in the ark: and God made a wind to pass over the earth, and the waters asswaged; The fountains also of the deep and the windows of heaven were stopped, and the rain from heaven was restrained.**"[88]

Not only did God **bring** the mighty storm in the first place, but He stopped it! He **caused** a wind to push back the waters; He **shut off** the fountains of the deep; He **stopped** the windows of heaven; and He **restrained** the rain. That amounts to excellent control over earth's weather systems.

Let me emphasize. This flood didn't just happen. Moses' account makes it very clear that **God did this**. The whole event was under God's control — it was not random nor accidental. He meant to do it at that particular time, and the two chapters in Genesis give the details.

How big was the flood? It covered the whole land area of the earth. "**And the waters prevailed exceedingly upon the earth; and all the high hills, that were under the whole heaven, were covered. Fifteen cubits upward did the waters prevail; and the mountains were covered.**"[89]

When the deluge event was finally over, God said, "**While the earth remaineth, seedtime and harvest, and cold and heat, and summer and winter, and day and night shall not cease.**"[90] This suggests that the skies cleared of the mist that covered them before the flood to reveal day and night, sunlight and darkness, and seasons as we know them. The creation of sunlight and moonlight occurred years earlier as recorded in Genesis 1, but it is entirely possible that they were not visible (or easily visible) through the mist until after the flood. Maybe. Maybe not. Either way, God introduced these new conditions to Adam and his family.

---

[88] Genesis 8:1-2

[89] Genesis 7:19-20

[90] Genesis 8:22

And then finally, God set the rainbow in the clouds as a sign to Himself and to mankind. "**I do set my bow in the cloud, and it shall be for a token of a covenant between Me and the earth. And it shall come to pass, when I bring a cloud over the earth, that the bow shall be seen in the cloud: And I will remember my covenant, which is between Me and you and every living creature of all flesh; and the waters shall no more become a flood to destroy all flesh. And the bow shall be in the cloud; and I will look upon it, that I may remember the everlasting covenant between God and every living creature of all flesh that is upon the earth. And God said unto Noah, 'This is the token of the covenant, which I have established between Me and all flesh that is upon the earth.'**"[91]

Everyone today is familiar with rainbows. Until the time of the flood, these verses suggest that no rainbows existed. They were new in the days after the flood. What is needed to produce a rainbow? The sun must be out and there must be clouds or rain in the sky. God said, "**I do set** my bow in the cloud." God did it. He formed the rainbows.

The covenant He established says that whenever He sees the rainbow in the skies, He will remember that He promised not to flood the whole earth again. And true to His word, there have been no new worldwide floods like the one of Noah's day. Lots of smaller floods and lots of rainbows have occurred, but no all-consuming floods.

The verses that deal with the great flood show that God had complete control over earth's weather: the rain, the winds, the fountains of the deep, and the clouds. We can explain from our understanding of physics how rainbows appear, but God set the conditions on earth in Noah's day for that phenomenon — rainbows — to occur.

The Famine in Joseph's Day

Near the end of Genesis, Moses told the story of Joseph, who was sold into slavery in Egypt by his jealous brothers. After arriving in Egypt, he prospered, becoming the Pharaoh's right-hand man. During the time of his service, Pharaoh had dreams. God interpreted Pharaoh's dreams to Joseph, who in turn explained the dreams to Pharaoh. The dreams showed that Egypt

---

[91] Genesis 9:13-17

was about to experience seven years of plenty, followed by seven years of famine which would cover all the land.

In these first two passages, Joseph told Pharaoh that God could interpret his dreams. "**And Pharaoh said unto Joseph, 'I have dreamed a dream, and there is none that can interpret it: and I have heard say of thee, that thou canst understand a dream to interpret it.' And Joseph answered Pharaoh, saying, 'It is not in me: God shall give Pharaoh an answer of peace.'**"[92] "**And Joseph said unto Pharaoh, 'The dream of Pharaoh is one: God hath shewed Pharaoh what He is about to do.'**"[93]

Notice that Joseph told Pharaoh the interpretations of the dreams were to show Pharaoh "**what He [God] is about to do.**" Seven years of plenty followed by seven years of famine were going to happen under God's control on this Pharaoh's watch. This wasn't just happenstance and God was not just predicting the future — He was giving Pharaoh fair warning to take advantage of the situation and prepare his land for it. **God was about to cause years of plenty, followed by years of dearth**, and He was warning Pharaoh in advance.

After Joseph reiterated that the information was from God, he interpreted the dream for Pharaoh. "**This is the thing which I have spoken unto Pharaoh: What God is about to do He sheweth unto Pharaoh. Behold, there come seven years of great plenty throughout all the land of Egypt: And there shall arise after them seven years of famine; and all the plenty shall be forgotten in the land of Egypt; and the famine shall consume the land; And the plenty shall not be known in the land by reason of that famine following; for it shall be very grievous.**"[94]

The famine was to be "**very grievous.**" Pharaoh needed to prepare. And he did. As a result of this information, Pharaoh placed Joseph over all of Pharaoh's people and over all the land of Egypt. "**And Pharaoh said unto Joseph, 'Forasmuch as God hath shewed thee all this, there is none so discreet and wise as thou art: thou shalt be over my house, and according unto thy word shall all my people be ruled: only in the throne will I be**

---

[92] Genesis 41:15-16

[93] Genesis 41:25

[94] Genesis 41:28-31

greater than thou.' And Pharaoh said unto Joseph, 'See, I have set thee over all the land of Egypt.'"[95]

The seven years of plenty came and went as predicted. "**And in the seven plenteous years the earth brought forth by handfuls.**"[96] God controlled the weather and caused all the crops to flourish. But then, as promised, the seven good years were followed by "**seven years of dearth.**" "**And the seven years of plenteousness, that was in the land of Egypt, were ended. And the seven years of dearth began to come, according as Joseph had said: and the dearth was in all lands; but in all the land of Egypt there was bread. And when all the land of Egypt was famished, the people cried to Pharaoh for bread: and Pharaoh said unto all the Egyptians, 'Go unto Joseph; what he saith to you, do.' And the famine was over all the face of the earth: And Joseph opened all the storehouses, and sold unto the Egyptians; and the famine waxed sore in the land of Egypt. And all countries came into Egypt to Joseph for to buy corn; because that the famine was so sore in all lands.**"[97]

According to this passage, "**the famine was over all the face of the earth,**" and "**all countries came into Egypt to Joseph for to buy corn.**" God had set this up so Pharaoh and all his people would have sufficient food to last through the seven years of famine. Through God's forewarning and Joseph's preparations, Egypt had such an excess saved up that people came from other countries to buy food during the famine.

At the end of the story, Joseph explained to his brothers why God had sent him into Egypt many years earlier. "**Now therefore be not grieved, nor angry with yourselves, that ye sold me hither: for <u>God did send me before you to preserve life</u>.**"[98] The early story was that Joseph's brothers sold him into slavery. But here we learn that God sent him into Egypt "**to preserve life.**" God used Joseph to save the lives of many people of many countries during this world-wide famine.

---

[95] Genesis 41:39-41

[96] Genesis 41:47

[97] Genesis 41:53-57

[98] Genesis 45:5

How does one cause seven years of plenty and seven years of famine? One controls the weather: the rain, the clouds, the winds, and the sunlight. God's ability to bring years of plenty and years of famine to this region occurred because God controls the wind and the rain.

Miraculous Events Leading to the Exodus

Years later when the Israelites were enslaved in Egypt under a new and different Pharaoh who didn't know Joseph, God had to perform miracles before Pharaoh to prove His identity.

Among those miracles, He turned "**all the waters that were in the river**" "**to blood.**"[99] He caused it "**to rain a very grievous hail**" on the people and the land.[100] Moses described the storm: "**and <u>the LORD sent thunder and hail</u>, and the fire ran along upon the ground; and <u>the LORD rained hail upon the land of Egypt</u>. So there was hail, and fire mingled with the hail, very grievous, such as there was none like it in all the land of Egypt since it became a nation.**"[101]

There had never been such a storm — rain, thunder, lightning, hail — very grievous — in the history of the Egyptians. All of this came from God: "**the Lord <u>sent</u> thunder and hail,**" and "**the Lord <u>rained</u> hail upon the land.**" It didn't just happen — God caused it to happen.

To demonstrate further that the signs were from the Lord, the Lord gave Moses the signal — and at the signal the storm would cease! Why? So Pharaoh would know (1) that it was from the Lord, and (2) that the earth is the Lord's. "**And Moses said unto him [Pharaoh], 'As soon as I am gone out of the city, I will spread abroad my hands unto the LORD; and <u>the thunder shall cease</u>, <u>neither shall there be any more hail</u>; <u>that thou mayest know how that the earth is the LORD'S.</u>**"[102]

---

[99] Exodus 7:20

[100] Exodus 9:18

[101] Exodus 9:23-24

[102] Exodus 9:29

"And Moses went out of the city from Pharaoh, and spread abroad his hands unto the LORD: and the thunders and hail ceased, and the rain was not poured upon the earth."[103]

Moses showed these signs to Pharaoh so he would know "**that the earth is the Lord's.**" Moses went out, lifted up his hands to the Lord, and the Lord caused the storm to cease. No more lightning! No more thunder! No more rain! No more hail! It all stopped — right then!

This is more clear evidence that God has complete command over the weather. Was this an isolated instance of God's supposed control over weather, rain, and wind? No. More was to come.

The Exodus — Crossing the Red Sea on Dry Ground

When Moses finally led the Israelites out of the land of Egypt, they were pursued by the Egyptian army. When they came to the Red Sea, God had a plan which allowed the people to walk across the sea bottom on dry ground. Moses gave the signal: he stretched out his hand over the sea and the Lord took over from there.

"And Moses stretched out his hand over the sea; and the LORD caused the sea to go back by a strong east wind all that night, and made the sea dry land, and the waters were divided."[104]

Moses didn't cause the sea to split apart — God did! The Lord used "**a strong east wind all that night**" to divide the waters and make the sea dry land. God had it under control.

The Israelites crossed the sea on dry ground, and when they reached the other side, Moses stretched forth his hand again, and God restored the sea. The only problem for the Egyptian army was that they had followed the Israelites through the sea. Whereas the Israelites crossed on dry land, the Egyptian army became bogged down by soupy, muddy ground that made it tough going for their chariots and horses. And When God caused the waters to return, the Egyptian army was stuck in the midst of the sea. All of the Egyptian army were lost! But the Israelites were safely across, standing on the eastern shore of the sea.

---

[103] Exodus 9:33

[104] Exodus 14:21

"And Moses stretched forth his hand over the sea, and the sea returned to his strength when the morning appeared; and the Egyptians fled against it; and the LORD overthrew the Egyptians in the midst of the sea."[105]

God rescued the Israelites from the Egyptians. The Egyptian armies, caught in the middle of the sea when it closed, were drowned. There was no escape for them.

Following the rescue from Egypt, the Israelites sang a song to celebrate God's great victory over the Egyptian army as well as their rescue from the Egyptians. This song is in Exodus 15. Part of this song follows:

"Thy right hand, O LORD, is become glorious in power: Thy right hand, O LORD, hath dashed in pieces the enemy. And in the greatness of Thine excellency Thou hast overthrown them that rose up against Thee: Thou sentest forth Thy wrath, which consumed them as stubble. And with the blast of Thy nostrils the waters were gathered together, the floods stood upright as an heap, and the depths were congealed in the heart of the sea. The enemy said, 'I will pursue, I will overtake, I will divide the spoil; my lust shall be satisfied upon them; I will draw my sword, my hand shall destroy them.' Thou didst blow with Thy wind, the sea covered them: they sank as lead in the mighty waters. Who is like unto Thee, O LORD, among the gods? Who is like Thee, glorious in holiness, fearful in praises, doing wonders? Thou stretchedst out Thy right hand, the earth swallowed them."[106]

Note that the Israelites gave all the credit to God, and note further the way they explained it: Moses led the people out of Egypt, but God was the power behind it. God was the Commander! God was the victor! God overthrew the Egyptians by sending forth His wrath. **"With the blast of Thy nostrils the waters were gathered together, the floods stood upright as an heap, and the depths were congealed in the heart of the sea." "Thou didst blow with Thy wind." "Thou stretchedst out Thy right hand, the earth swallowed them."** God used His nostrils to gather the waters; He blew with His wind; and the sea and earth swallowed them!

---

[105] Exodus 14.27

[106] Exodus 15:6-12

These events show that God exerted complete control over the earth, the wind, and the sea to rescue the Israelites from the land of Egypt and to destroy the Egyptian armies pursuing them.

In addition to this event, while the Israelites were wandering in the desert[107], God made bitter water sweet for them to drink. Then, He gave them quail and manna to eat[108]. Later, God told Moses to smite the rock, and water flowed forth for the people to drink[109]. God continued to exercise control over their living conditions as He guided the Israelites through the wilderness towards the Promised Land.

The Promised Land

Leading to their entry into the Promised Land, God told the Israelites that if they loved the Lord and served Him, He would send rain to make their crops bring forth abundantly. If they did not properly reverence the Lord, He would withhold the rain, the land would not give fruit, and they would perish from the land. Here again, God showed that He had control over the wind and the rain and over His people living in the land.

"But the land, whither ye go to possess it, is a land of hills and valleys, and drinketh water of the rain of heaven: A land which the LORD thy God careth for: the eyes of the LORD thy God are always upon it, from the beginning of the year even unto the end of the year. And it shall come to pass, if ye shall hearken diligently Unto My commandments which I command you this day, to love the LORD your God, and to serve Him with all your heart and with all your soul, That I will give you the rain of your land in his due season, the first rain and the latter rain, that thou mayest gather in thy corn, and thy wine, and thine oil. And I will send grass in thy fields for thy cattle, that thou mayest eat and be full. Take heed to yourselves, that your heart be not deceived, and ye turn aside, and serve other gods, and worship them; And then the LORD'S wrath be kindled against you, and He shut up the heaven, that there be no rain, and that the

---

[107] Exodus 15:25

[108] Exodus 16:13-15

[109] Exodus 17.6

land yield not her fruit; and lest ye perish quickly from off the good land which the LORD giveth you."[110]

Later in Deuteronomy, the Lord included the giving of rain from heaven as part of "**His good treasure**" which He was going to provide to the Israelites. "**And the LORD shall make thee plenteous in goods, in the fruit of thy body, and in the fruit of thy cattle, and in the fruit of thy ground, in the land which the LORD sware unto thy fathers to give thee. The LORD shall open unto thee His good treasure, the heaven to give the rain unto thy land in His season, and to bless all the work of thine hand: and thou shalt lend unto many nations, and thou shalt not borrow.**"[111]

The Lord promised them much more in this passage than just good weather. He promised other benefits of His power to include the fruit of their bodies, the fruit of their cattle, and the fruit of the ground. The Lord has much broader control over His creation than just the wind and the rain. The power to make His people and their animals fruitful is much more important than His ability to control wind and rain. But rain is necessary for man to drink and for food to grow — and the Lord was going to provide all of that to the people. He promised to make the Israelites prosper in the land. He would make them "**plenteous in goods,**" plenteous "**in the fruit of thy ground,**" and He would bless the work of their hands.

The rain and weather were God's to control and He chose to bless them with both which were necessary for their success in the land. He promised rain in abundance and the opening of "**His good treasure**" for their benefit.

Israelites vs Amorites

God used hailstones against the Amorites. "**And it came to pass, as they fled from before Israel, and were in the going down to Bethhoron, that the LORD cast down great stones from heaven upon them unto Azekah,**

---

[110] Deuteronomy 11:11-17

[111] Deuteronomy 28:11-12

and they died: they were more which died with hailstones than they whom the children of Israel slew with the sword."[112]

God used His control over the heaven and the earth, and in particular over the wind and the rain, to cast large hailstones upon these enemies of the children of Israel, the Amorites. Once again, it is clear that these "**great stones from heaven**" did not just happen to fall. God caused them to fall in that place at that time.

Hail comes from rain passing through cold air masses which cause the rain droplets to freeze. Large hail stones come from strong winds that throw the small stones back up through the moisture (where they are coated again with water) and through the cold air masses where the new layers freeze. The stronger the winds, the bigger the size of hailstones that can be blown back up into the clouds for another pass and for more growth. That is why the most violent thunderstorms can contain the largest hailstones.

Here again, God demonstrated His power over the wind and the rain.

Elijah's Prophecy

The prophet Elijah told Ahab that the Lord was about to withhold the rain for several years. Elijah did not have any intrinsic power to perform such a miracle, but as a prophet of God, he relayed the information from God to Ahab. God withheld the rain.

"**And Elijah the Tishbite, who was of the inhabitants of Gilead, said unto Ahab, 'As the LORD God of Israel liveth, before whom I stand, there shall not be dew nor rain these years, but according to my word.'"**[113]

A few verses later, Elijah quoted the Lord directly: "**For thus saith the LORD God of Israel, 'The barrel of meal shall not waste, neither shall the cruse of oil fail, until the day that the LORD sendeth rain upon the earth.'"**[114]

God was going to withhold the rain from the earth, and it would not rain again until God decided. After the long period without rain, the Lord sent

---

[112] Joshua 10:11

[113] 1 Kings 17:1

[114] 1 Kings 17:14

Elijah to tell Ahab that He [God] was ready once again to send the rains. **"And it came to pass after many days, that the word of the LORD came to Elijah in the third year, saying, 'Go, shew thyself unto Ahab; and I will send rain upon the earth.'"**[115]

Most of Chapter 18 of 1 Kings describes God's demonstration to prove to Ahab and to the people that He, and not Baal, was truly God. At the end of the chapter and after these signs, the people once again bowed down to worship God — and God once again favored them with rain.

"And Elijah said unto Ahab, 'Get thee up, eat and drink; for there is a sound of abundance of rain.'"[116] **"And it came to pass in the mean while, that the heaven was black with clouds and wind, and there was a great rain. And Ahab rode, and went to Jezreel."**[117]

All the while the people were worshiping Baal, God withheld the rain. But after they bowed down to God Himself, He brought back dark storm clouds, wind, and rain. Elijah was God's spokesman. God did the work. God withheld the rain, and then supplied it again. Since God controls the wind and the rain, no one should be surprised.

Elisha's Prophecy

The kings of Edom, Israel, and Judah called for the Prophet Elisha to help them enquire of the Lord. The hand of the Lord came upon Elisha and he relayed the following message to them.

"For thus saith the LORD, 'Ye shall not see wind, neither shall ye see rain; yet that valley shall be filled with water, that ye may drink, both ye, and your cattle, and your beasts.' And this is but a light thing in the sight of the LORD: He will deliver the Moabites also into your hand. And ye shall smite every fenced city, and every choice city, and shall fell every good tree, and stop all wells of water, and mar every good piece of land with stones. And it came to pass in the morning, when the meat offering was

---

[115] 1 Kings 18:1

[116] 1 Kings 18:41

[117] 1 Kings 18:45

offered, that, behold, there came water by the way of Edom, and the country was filled with water."[118]

The Lord said He was not going to bring wind or rain upon their land, but He was nevertheless going to fill the valley with water. And by verse 20, He had done so. Here again He showed that He was in control of the wind, the rain, and in this case also, the water.

Solomon's Request

In 2 Chronicles, Solomon prayed for God to forgive His people Israel because the people had sinned and God had withheld the rain. Solomon asked God to forgive their sins and send rain once again upon the land if the people would repent.

"When the heaven is shut up, and there is no rain, because they have sinned against Thee; yet if they pray toward this place, and confess Thy name, and turn from their sin, when Thou dost afflict them; then hear Thou from heaven, and forgive the sin of Thy servants, and of Thy people Israel, when Thou hast taught them the good way, wherein they should walk; and send rain upon Thy land, which Thou hast given unto Thy people for an inheritance."[119]

This shows that Solomon and the children of Israel understood clearly that control over the rain was in God's hand. He had afflicted them with no rain, and they knew they had to request rain again directly from Him. God's answer to Solomon followed in the next chapter.

"And the LORD appeared to Solomon by night, and said unto him, 'I have heard thy prayer, and have chosen this place to Myself for an house of sacrifice. If I shut up heaven that there be no rain, or if I command the locusts to devour the land, or if I send pestilence among My people; If My people, which are called by My name, shall humble themselves, and pray, and seek My face, and turn from their wicked ways; then will I hear from heaven, and will forgive their sin, and will heal their land."[120]

---

[118] 2 Kings 3:17-20

[119] 2 Chronicles 6:26-27

[120] 2 Chronicles 7:12-14

Here, God admitted to having control over the rain, over the locusts, and over pestilence. He admitted His control extends beyond the wind and the rain in this case. But after having shut up heaven to withhold the rain, if the people were to repent, God said He would do as requested and "**heal their land.**" That is, if they repented, He would send rain once again.

God controls the wind and the rain, and Solomon knew it.

God's Dealings with Job

Job was a man of God. But God had given the devil license to test Job. As a result, Job lost everything and was suffering greatly. Eliphaz suggested to Job that he call upon God and commit his situation to God. According to Eliphaz, God, who performs great, wonderful, unsearchable, marvelous things without number, could certainly help. Among those things that God could do, Eliphaz included, "**who giveth rain upon the earth, and sendeth waters upon the fields.**"[121] Eliphaz recognized that God had control over the rain and the waters of the earth. He recognized God held the power and he suggested Job take his problems to God.

Job acknowledged God's powers: "**Behold, He withholdeth the waters, and they dry up: also He sendeth them out, and they overturn the earth.**"[122] "**He bindeth up the waters in His thick clouds; and the cloud is not rent under them. He holdeth back the face of His throne, and spreadeth His cloud upon it.**"[123]

Job continued: "**To make the weight for the winds; and He weigheth the waters by measure. When He made a decree for the rain, and a way for the lightning of the thunder: Then did He see it, and declare it; He prepared it, yea, and searched it out. And unto man He said, 'Behold, the fear of the Lord, that is wisdom; and to depart from evil is understanding.'**"[124]

---

[121] Job 5:10

[122] Job 12:15

[123] Job 26:9

[124] Job 28:25-28

Job recognized God's control over wind, rain, lightning, and thunder but he still lacked in understanding.

Elihu then reminded Job of God's attributes: "**For He maketh small the drops of water: they pour down rain according to the vapour thereof: Which the clouds do drop and distil upon man abundantly. Also can any understand the spreadings of the clouds, or the noise of His tabernacle? Behold, He spreadeth His light upon it, and covereth the bottom of the sea. For by them judgeth He the people; He giveth meat in abundance. With clouds He covereth the light; and commandeth it not to shine by the cloud that cometh betwixt. The noise thereof sheweth concerning it, the cattle also concerning the vapour.**"[125]

Job and his friends seemed to know a lot about God, but it does not appear that Job acted on any of that knowledge. In this next passage, Elihu spoke of God's control of the clouds and the rain. It is clear that Elihu and Job were well aware that God was in control of the wind, rain, and clouds.

Elihu continued: "**God thundereth marvellously with His voice; great things doeth He, which we cannot comprehend. For He saith to the snow, 'Be thou on the earth;' likewise to the small rain, and to the great rain of His strength. He sealeth up the hand of every man; that all men may know His work. Then the beasts go into dens, and remain in their places. Out of the south cometh the whirlwind: and cold out of the north. By the breath of God frost is given: and the breadth of the waters is straitened. Also by watering He wearieth the thick cloud: He scattereth His bright cloud: And it is turned round about by His counsels: that they may do whatsoever He commandeth them upon the face of the world in the earth. He causeth it to come, whether for correction, or for His land, or for mercy.**"

"**Hearken unto this, O Job: stand still, and consider the wondrous works of God. Dost thou know when God disposed them, and caused the light of His cloud to shine? Dost thou know the balancings of the clouds, the wondrous works of Him which is perfect in knowledge? How thy garments are warm, when He quieteth the earth by the south wind? Hast thou with Him spread out the sky, which is strong, and as a molten looking glass? Teach us what we shall say unto Him; for we cannot order our**

---

[125] Job 36.27-33

speech by reason of darkness. Shall it be told Him that I speak? if a man speak, surely he shall be swallowed up. And now men see not the bright light which is in the clouds: but the wind passeth, and cleanseth them."[126]

Elihu showed in this passage that God controls the rain, the snow, the frost, the warm south winds, and the cold north winds, although he could not explain how God does it. He nevertheless was pointing Job in the right direction — towards God.

Finally, God spoke directly to Job "**out of the whirlwind**" in Job 38. Actually, He confronted Job! "Who do you think you are??!! Where were you when I created the world??!! Explain it to Me if you think you can!!" The questions come across as really indignant outpourings from a riled God who was dealing with an arrogant man by the name of Job. But with this confrontation, God put him in his place!

"Then the LORD answered Job out of the whirlwind, and said, 'Who is this that darkeneth counsel by words without knowledge? Gird up now thy loins like a man; for I will demand of thee, and answer thou Me. <u>Where wast thou when I laid the foundations of the earth? Declare, if thou hast understanding</u>. Who hath laid the measures thereof, if thou knowest? Or who hath stretched the line upon it? Whereupon are the foundations thereof fastened? Or who laid the corner stone thereof.'"[127]

This is only the beginning of God's questions to Job — but look at them. "Who darkens counsel without knowledge? Where were you when I created the earth? Tell me if you have understanding!! Who measured the earth?? Who laid its foundations?? Who laid its corner stone?? Tell me if you know!!"

Surely in this passage, God was talking specifically to Job. But the questions from God apply to **everyone** who thinks he knows God, who thinks he understands God's creation, or who thinks he knows how to control God's creation! How many of us are included in the broader applications of these questions? EVERYONE!! We all are!!

God continued: "**Hast thou entered into <u>the treasures of the snow</u>? Or hast thou seen the treasures of <u>the hail</u>, which I have reserved against the time of trouble, against the day of battle and war? By what way is <u>the</u>**

---

[126] Job 37:3-21

[127] Job 38.1-6

light parted, which scattereth the east wind upon the earth? Who hath divided a watercourse for the overflowing of waters, or a way for the lightning of thunder; To cause it to rain on the earth, where no man is; on the wilderness, wherein there is no man; To satisfy the desolate and waste ground; and to cause the bud of the tender herb to spring forth? Hath the rain a father? or who hath begotten the drops of dew? Out of whose womb came the ice? and the hoary frost of heaven, who hath gendered it? The waters are hid as with a stone, and the face of the deep is frozen."[128]

God asked Job to explain, all of this — any of this — if he could. "Who controls the wind, rain, hail, lightning, dew, frost, and the waters? Who decides when they each are needed? Who understands the details of each of them?" Job had no idea. Men today have no idea, either.

Yes, we know that God controls all of this, but we have no idea how He controls, when and how He applies each of these phenomena, or how He produces any of them. And He still wasn't finished with His questions.

"Can you [Job] speak to the clouds and bring water? Can you send lightning? Can you send clouds, or count the clouds, or stay the clouds? Who controls the weather over the earth?? Can you do any of this?"

"Canst thou lift up thy voice to the clouds, that abundance of waters may cover thee? Canst thou send lightnings, that they may go, and say unto thee, 'Here we are?' Who hath put wisdom in the inward parts? or who hath given understanding to the heart? Who can number the clouds in wisdom? or who can stay the bottles of heaven."[129]

Job knew the answers to all of these questions: "You can, God! You know, God!" But there was still more to come: "Are you, Job, going to teach me? Is mankind going to teach God??!! Are you able to stand before Me?? This is MY LAND! This is GOD's LAND!"

"Shall he that contendeth with the Almighty instruct Him? He that reproveth God, let him answer it."[130] **"None is so fierce that dare stir him up: who then is able to stand before Me? Who hath prevented Me,**

---

[128] Job 38:22-30

[129] Job 38:34-37

[130] Job 40:2

that I should repay him? Whatsoever is under the whole heaven is Mine."[131]

Job thought he was smart, capable, and knowledgeable, but when confronted by an angry God, he said nothing. **"Then Job answered the LORD, and said, 'Behold, I am vile; what shall I answer Thee? I will lay mine hand upon my mouth. Once have I spoken; but I will not answer: yea, twice; but I will proceed no further.'"**[132] Job must have thought, "I spoke once but no more. I will keep my mouth shut."

God created the heavens and the earth. And God controls them. God controls the wind and the rain and all of the various types of weather. Can we [mankind] control any of this by our own [mankind's] power? ("Mankind's power" — sounds like an oxymoron!) NO, we can't control any of this!! Can we even understand it? Same answer — NO!

Herein lies man's problem. Mankind, like Job, think they can understand God. They can duplicate His abilities, they can understand His ways, they can predict His weather, and they can alter the long-range climate which He controls. Yet the majority of mankind want nothing to do with God. They, like Job, believe man is more capable than he actually is.

God is in control of the wind, rain, storms, lightning, hail, etc., ... the weather. He made that point very clear to Job!

But this isn't the last mention of God's control of the wind and the rain in the Bible. More evidence continues.

The Psalms

David called upon the Lord in Psalm 18. In this psalm, note all of the phenomena he attributed to the Lord: the earth shook and trembled; the foundations were moved and shaken; smoke went out of His nostrils; fire (lightning??) went out of His mouth; He flew upon the wind; darkness, dark waters, and thick clouds were about Him; hail stones and coals of fire passed from His thick clouds; and thunder, arrows (bolts of lightning??), and great winds came from the breath of His nostrils to accompany His rebuke. Only one

---

[131] Job 41:10-11

[132] Job 40:3-5

who controls the heavens, the earth, the winds, and the rains can perform such actions.

"In my distress I called upon the LORD, and cried unto my God: He heard my voice out of His temple, and my cry came before Him, even into His ears. Then the earth shook and trembled; the foundations also of the hills moved and were shaken, because He was wroth. There went up a smoke out of His nostrils, and fire out of His mouth devoured: coals were kindled by it. He bowed the heavens also, and came down: and darkness was under His feet. And He rode upon a cherub, and did fly: yea, He did fly upon the wings of the wind. He made darkness His secret place; His pavilion round about Him were dark waters and thick clouds of the skies. At the brightness that was before Him His thick clouds passed, hail stones and coals of fire. The LORD also thundered in the heavens, and the Highest gave His voice; hail stones and coals of fire. Yea, He sent out His arrows, and scattered them; and He shot out lightnings, and discomfited them. Then the channels of waters were seen, and the foundations of the world were discovered at Thy rebuke, O LORD, at the blast of the breath of Thy nostrils."[133]

Following this passage, David gave a general statement about the Lord's relationship to the world. **"The earth is the LORD'S, and the fulness thereof; the world, and they that dwell therein. For He hath founded it upon the seas, and established it upon the floods."**[134] The earth, the world, the seas, and the floods are His to command. And all lives that dwell therein are the Lord's.

The waters of the seas are central to God's creation. His voice is upon the waters. **"The voice of the LORD is upon the waters: the God of glory thundereth: the LORD is upon many waters."**[135]

He controls the waters. **"He gathereth the waters of the sea together as an heap: He layeth up the depth in storehouses."**[136]

---

[133] Psalms 18:6-15

[134] Psalms 24:1-2

[135] Psalms 29:3

[136] Psalms 33:7

The Psalmist praised God for His control over the waters: "**Thou visitest the earth, and waterest it: Thou greatly enrichest it with the river of God, which is full of water: Thou preparest them corn, when Thou hast so provided for it. Thou waterest the ridges thereof abundantly: Thou settlest the furrows thereof: Thou makest it soft with showers: Thou blessest the springing thereof.**"[137]

... and for His control over the rain "**Thou, O God, didst send a plentiful rain, whereby Thou didst confirm Thine inheritance, when it was weary.**"[138]

... and for His control over the sea. "**Thou rulest the raging of the sea: when the waves thereof arise, Thou stillest them.**"[139]

... and for His control over the lightning. "**His lightnings enlightened the world: the earth saw, and trembled.** "[140]

... and for His control over storms and waves. "**These see the works of the LORD, and His wonders in the deep. For He commandeth, and raiseth the stormy wind, which lifteth up the waves thereof.**"[141] "**He maketh the storm a calm, so that the waves thereof are still.**"[142]

... and for His control over the rivers and waters of the earth. "**He turneth rivers into a wilderness, and the watersprings into dry ground; A fruitful land into barrenness, for the wickedness of them that dwell therein. He turneth the wilderness into a standing water, and dry ground into watersprings.**"[143]

---

[137] Psalms 65:9-10

[138] Psalms 68:9

[139] Psalms 89:9

[140] Psalms 97:4

[141] Psalms 107:24-25

[142] Psalms 107:29

[143] Psalms 107:33-35

Whatever the Lord wants to do, He does. "**Whatsoever the LORD pleased, that did He in heaven, and in earth, in the seas, and all deep places. He causeth the vapours to ascend from the ends of the earth; He maketh lightnings for the rain; He bringeth the wind out of His treasuries.**"[144]

Further praises are given to the Lord for the snow, frost, ice, temperature, and wind. In other words, He receives lots of praises for His overall control of the weather. "**He giveth snow like wool: He scattereth the hoarfrost like ashes. He casteth forth His ice like morsels: Who can stand before His cold? He sendeth out His word, and melteth them: He causeth His wind to blow, and the waters flow.**"[145]

None of these phenomena just happen in and of themselves — God causes them and God controls them.

Wisdom Speaks

Wisdom speaks in Proverbs 8. It is the "I" in that passage. It addressed subjects appropriate to the rain and the wind. "**When He prepared the heavens, I was there: when He set a compass upon the face of the depth: When He established the clouds above: when He strengthened the fountains of the deep: When He gave to the sea His decree, that the waters should not pass His commandment: when He appointed the foundations of the earth: Then I was by Him, as one brought up with Him: and I was daily His delight, rejoicing always before Him; Rejoicing in the habitable part of His earth; and my delights were with the sons of men.**"[146]

God created all things and prepared the heavens, the clouds above, the fountains of the deep, the sea, the waters, and earth's foundations **with wisdom**. "I," said Wisdom, "was there." All of His creation was formed using the wisdom of God. It is a wonderful creation, and He still controls it.

---

[144] Psalms 135:6-7

[145] Psalms 147:16-18

[146] Proverbs 8:27-31

Circuits and Cycles

God set up the earth with the following circuits: The sun rises and sets as the earth spins round and round on its axis; the winds flow along certain known circuits continually; and the rain, which falls upon the earth, forms rivers that run into the sea, where the water evaporates to form clouds which produce rain once again.

"The sun also ariseth, and the sun goeth down, and hasteth to his place where he arose. The wind goeth toward the south, and turneth about unto the north; it whirleth about continually, and the wind returneth again according to his circuits. All the rivers run into the sea; yet the sea is not full; unto the place from whence the rivers come, thither they return again."[147]

Ancient mariners knew of the latitudes that produced westward and eastward prevailing winds. Depending where they wanted to sail, they first sailed north or south to the appropriate latitude and then they could catch the prevailing wind to take them to their destinations across the ocean. The sea itself contains streams (the Gulf Stream, for example) that flow in circuits as well. There are many known cycles in this earth — such as the water cycle mentioned above. God set up each of these circuits and cycles to keep the earth functioning properly. He also has the ability to alter and control these cycles.

Jeremiah's Prophecies

Jeremiah spoke concerning the children of Israel: **"Neither say they in their heart, <u>Let us now fear the LORD our God, that giveth rain, both the former and the latter, in His season</u>: He reserveth unto us the appointed weeks of the harvest."**[148] He accused the children of Israel of turning away from God and not remembering that God controls the rain and the success or failure of their crops.

---

[147] Ecclesiastes 1:5-7

[148] Jeremiah 5:24

The children of Israel should have been very familiar with God and His capabilities. But regardless how many times they were reminded, they had mostly forgotten who God was and what He could do.

Jeremiah warned them to remember. "**But the LORD is the true God, He is the living God, and an everlasting king: at His wrath the earth shall tremble, and the nations shall not be able to abide His indignation.**"[149] "**He hath made the earth by His power, He hath established the world by His wisdom, and hath stretched out the heavens by His discretion. When He uttereth His voice, there is a multitude of waters in the heavens, and He causeth the vapours to ascend from the ends of the earth; He maketh lightnings with rain, and bringeth forth the wind out of His treasures.**"[150]

Twice Jeremiah reminded the children of Israel that they were dealing with the one true God, who made the earth, the world, and the heavens with His power, wisdom, and understanding. He also warned them that this same God controls the waters, vapors, lightning, rains, and winds.

A little later, Jeremiah was even more direct when he told them that "**they have forsaken the Lord, the fountain of living waters.**"[151] This title clearly suggests that the Lord controls life and death because He controls the waters, i.e., He is the "**fountain of living waters.**"

Ezekiel's Prophecies

Ezekiel confronted the false prophets of the land, who were leading the children of Israel astray, with God's words. "**And Mine hand shall be upon the prophets that see vanity, and that divine lies: they shall not be in the assembly of My people, neither shall they be written in the writing of the house of Israel, neither shall they enter into the land of Israel; and ye shall know that I am the Lord GOD. Because, even because they have seduced My people, saying, 'Peace;' and there was no peace; and one built up a wall, and, lo, others daubed it with untempered morter: Say unto them which daub it with untempered morter, that it shall fall: there shall be an**

---

[149] Jeremiah 10:10

[150] Jeremiah 10:12-13 and 51:15-16

[151] Jeremiah 17:13

overflowing shower; and ye, O great hailstones, shall fall; and a stormy wind shall rend it. Lo, when the wall is fallen, shall it not be said unto you, Where is the daubing wherewith ye have daubed it? Therefore thus saith the Lord GOD; I will even rend it with a stormy wind in my fury; and there shall be an overflowing shower in mine anger, and great hailstones in my fury to consume it."[152]

The false prophets were lying to the people. They were telling the people that God was promising peaceful times, when God was actually warning of terrible times to come. Some were promising strong walls formed by adding untempered mortar to inferior quality walls. The author has seen walls like this in the Far East. They look wonderfully sturdy, but they are not. Leave most of the cement out of stucco, and it forms a wall that looks wonderful, but which you can easily punch a fist through.

The lie was that this protective wall would be sturdy — but it was not. These prophets were telling the people they would be safe — when there was no safety at all. God was going to show them the true nature of this wall as He brought "**an overflowing shower**," "**great hailstones**," and "**a stormy wind**" to break it apart and cause it to fall. Certainly, a good wall should stand up to the wind, rain, and hail. But the wind, shower, and hailstones, which were sent directly by God in His fury and anger, would consume the wall. Actually, very few walls, even well-built ones can stand up to the power and might of a strong thunderstorm. This shows the power of the wind, the rain, and the hailstones which God had at his beck and call.

Think of the destructive power of hurricanes and tornados. These, too, result from the wind and the rain which God controls.

Ezekiel also warned the people of the end times which would bring great tribulation upon the land. When "**Gog**" comes up against the land of Israel, God will bring overflowing rain and great hailstones (along with fire and brimstone) upon him.

"And I will plead against him with pestilence and with blood; and I will rain upon him, and upon his bands, and upon the many people that are with him, an overflowing rain, and great hailstones, fire, and brimstone."[153]

---

[152] Ezekiel 13:9-13

[153] Ezekiel 38:22

Minor Prophecies

Amos told the children of Israel that God had withheld the rain from them. God said, "**And also I have withholden the rain from you, when there were yet three months to the harvest: and I caused it to rain upon one city, and caused it not to rain upon another city: one piece was rained upon, and the piece whereupon it rained not withered.**"[154] That God controlled the rain and wind did not seem to be any great shock to the children of Israel. It should not have been a shock to them — they should have known these phenomena were under God's control. In this case, God said He caused it to rain upon the land, and He withheld the rain, as He saw fit. The crops withered where He withheld the rain; and they prospered where He sent it.

More words of warning came from God through Amos. In this case, Amos simply warned that God had control over all waters — and the people should beware. "**And the Lord GOD of hosts is He that toucheth the land, and it shall melt, and all that dwell therein shall mourn: and it shall rise up wholly like a flood; and shall be drowned, as by the flood of Egypt. It is He that buildeth His stories in the heaven, and hath founded His troop in the earth; He that calleth for the waters of the sea, and poureth them out upon the face of the earth: The LORD is His name.**"[155]

According to Amos, God controlled the land and the sea, and poured the waters out upon the earth as He saw fit.

Nahum also warned the people when he described the actions of a jealous God: "**God is jealous, and the LORD revengeth; the LORD revengeth, and is furious; the LORD will take vengeance on His adversaries, and He reserveth wrath for His enemies. The LORD is slow to anger, and great in power, and will not at all acquit the wicked: the LORD hath His way in the whirlwind and in the storm, and the clouds are the dust of his feet. He rebuketh the sea, and maketh it dry, and drieth up all the rivers: Bashan languisheth, and Carmel, and the flower of Lebanon languisheth.**"[156]

---

[154] Amos 4:7

[155] Amos 9:5-6

[156] Nahum 1:2-4

In this case, he warned that God controls whirlwinds, storms, rivers, and the sea. This specific warning was that God would direct the whirlwind, the storm, and the clouds to dry up the sea and the rivers. They were His to control, and Nahum warned the people to watch out!

Zechariah told the people to ask the Lord for rain when it was needed. The Lord can give rain, make bright clouds, and give showers of rain. They needed only to look to God. "**Ask ye of the LORD rain in the time of the latter rain; so the LORD shall make bright clouds, and give them showers of rain, to every one grass in the field.**"[157] Why? If they wanted their crops to grow well, they needed to request rain from the Lord — because He controls the rain.

OT Summary

All throughout the Old Testament, we see that God has control over the wind, rain, clouds, storms, lightning, etc. That is, God controls the weather in all its forms. Over and over again, throughout the Old Testament, we see examples of His control over earth's weather systems. But the evidence doesn't end with the Old Testament.

Evidence of the fact that God controls the wind, rain, and weather continues in the New Testament as well.

Evidence in the Gospels

In the Gospels, Jesus commanded the rain, the winds, and the sea to obey Him. Specifically, He rebuked a violent storm and caused it to cease. Three of the four Gospel writers (Matthew, Mark, and Luke) recorded this one event. Those who saw it were amazed. Those who read about it should be amazed also. Remember, in the Gospels, we are looking at eye-witness accounts. Following is Mark's account of this event.

"**And the same day, when the even was come, He saith unto them, 'Let us pass over unto the other side.' And when they had sent away the multitude, they took Him even as He was in the ship. And there were also with Him other little ships. And there arose a great storm of wind, and the**

---

[157] Zechariah 10:1

waves beat into the ship, so that it was now full. And He was in the hinder part of the ship, asleep on a pillow: and they awake Him, and say unto Him, 'Master, carest Thou not that we perish?' And He arose, and rebuked the wind, and said unto the sea, 'Peace, be still.' And the wind ceased, and there was a great calm. And He said unto them, 'Why are ye so fearful? How is it that ye have no faith?' And they feared exceedingly, and said one to another, 'What manner of man is this, that even the wind and the sea obey Him?'"[158]

The scene shows Jesus and the men in a ship crossing the sea. A violent storm had come up. The men were apparently battling the storm, while the ship, nevertheless, was filling with water. Where was Jesus? ... asleep in the back of the boat. The storm was so fierce that they feared for their lives.

So, they woke Jesus. "Aren't you afraid?! We're going to die in this storm!" Jesus then rose up, rebuked the wind, and told the sea to be still! Immediately, there was a great calm. It was such a dramatic, instantaneous change that the men feared exceedingly. This time, they feared Jesus' power. They quickly questioned one another, "What manner of man is this?"

Luke's version showed the men questioning because He commanded, and the winds and water obeyed Him. **"And He said unto them, 'Where is your faith?' And they being afraid wondered, saying one to another, 'What manner of man is this! for He commandeth even the winds and water, and they obey Him.'"**[159]

This man was Jesus, The Creator — the Son of God. When He spoke, even the winds, water, sea, and the rain obeyed.

These New Testament passages confirm the veracity of the evidence from the Old Testament. In this case, however, the name of the man, Jesus, is included. We know very clearly from the New Testament Gospels that Jesus is the Son of God. We know very clearly from these passages that He controls the wind and the rain. He speaks — they obey!

James repeated the story about Elijah [Elias] praying to God to withhold the rain. And God withheld it! Then Elijah prayed again for rain, and God gave them rain.

---

[158] Mark 4:35-41

[159] Luke 8:22-25

"Elias was a man subject to like passions as we are, and he prayed earnestly that it might not rain: and it rained not on the earth by the space of three years and six months. And he prayed again, and the heaven gave rain, and the earth brought forth her fruit."[160]

Once again, note that Elijah did not withhold the rain. He prayed to God, who withheld the rain. Then he prayed again and God brought forth rain. God did the work. This New Testament passage confirms the veracity of the Old Testament story. What does it show? It shows once more that God controls the wind and the rain.

Finally, in the Book of Revelation, we see lightning, thunders, earthquakes, and great hail associated with God's temple. "**And the temple of God was opened in heaven, and there was seen in His temple the ark of His testament: and there were lightnings, and voices, and thunderings, and an earthquake, and great hail.**"[161]

Later, when the seventh angel poured out his vial of judgement into the earth's atmosphere, we see the same phenomena emanating from God's temple. "**And the seventh angel poured out his vial into the air; and there came a great voice out of the temple of heaven, from the throne, saying, 'It is done.' And there were voices, and thunders, and lightnings; and there was a great earthquake, such as was not since men were upon the earth, so mighty an earthquake, and so great.**"[162]

The final storm we see in this passage is a storm that brings great hail upon mankind. "**And there fell upon men a great hail out of heaven, every stone about the weight of a talent: and men blasphemed God because of the plague of the hail; for the plague thereof was exceeding great.**"[163]

This passage indicates once again that mankind was aware of the source of this terrible storm and the great hailstones. They recognized that the storm and hailstones came from God. They recognized that it came upon them as judgement, but they nevertheless "**blasphemed God because of the plague**

---

[160] James 5:17-18

[161] Revelation 11:19

[162] Revelation 16:17-18

[163] Revelation 16:21

of the hail." They recognized God was the source — and they blasphemed God. Not a lot of respect will be shown to God in that coming day. Not a lot of respect was shown to God in Old Testament times, and not a lot of respect is shown to Him today. Someday, He will receive the respect He is due, but that day isn't here yet.

Until then, we can count on Him to continue to control the wind, the rain, and the storms as He wishes.

Summary

Evidence throughout the Bible shows that the same God who created and maintains the heavens and the earth also controls the wind, rain, lightning, storms, and hail. That is, the Creator God controls the weather and the climate. The biblical evidence overwhelmingly shows that God Himself can and does control the wind and the rain as He sees fit.

If the only mention of God **creating** the heavens and the earth, or **controlling** the heavens and the earth, or **controlling the wind and the rain** occurred in Genesis, written by Moses, one might argue that Moses' witness was faulty. A lone testimony? Who would believe that? But the evidence of the answer to the question, **Who controls the wind and the rain?** is spread throughout the whole Bible. Evidence from more than twenty books of the Bible indicates that God controls them.

Rather than finding next to nothing about the world's climate in the Bible, we find many examples throughout the whole Bible which demonstrate that God is in control of the wind and the rain. That means God is in control of the climate and all forms of weather.

Does mankind take that into account in their thinking? Not really. Prior to this study, the author didn't anticipate the volume of evidence to be found. But now it is clear — the evidence is overwhelming!

5

The Bible vs The Copernican & Mediocrity Principles

Years ago, the church tangled with science. It saw an opportunity to take a scientific theory and turn it into a church doctrine. It was not necessary to their religious point of view, but the church adopted it anyway.

The exact location of the earth, relative to our sun and the other planets, has nothing to do with God's view of the earth and mankind. Both earth and mankind **are,** and always will be, special in His eyes.

Ptolemy

Claudius Ptolemy, a Roman citizen of the 2nd Century A.D., is credited with proposing the *geocentric* model of the universe in which earth occupies a stationary position at its center.[164] According to Ptolemy, the sun, all planets, and everything else in the universe revolve around the earth.

This geocentric theory was apparently too much for the church to pass up. It was such a great theory that the Roman Catholic church adopted it as a church doctrine. With this theory, science and the Bible both gave the earth and mankind prominence in the universe. With its adoption, both science and the church agreed that the earth and mankind held exalted positions in the universe.

---

[164] Wikipedia, "Geocentric Model."

But scientific discovery progressed and our understanding of our solar system changed. It advanced due to better telescopes and better, more precise, observational data.

Copernicus

About the 16th Century, Nicolaus Copernicus suggested the *heliocentric* model of the solar system. In his model, all planets in our solar system orbit the sun. The Roman Catholic church objected. Regardless of what the Bible said, they understood Copernicus' theory to suggest that the earth and mankind were not special in the grand scheme of the universe. This idea became known as *the Copernican Principle.*

Copernicus' theory came from studies of the relative motions of the earth, the sun, and the other planets. Were the earth truly at the center of the system, all planets as well as the sun would continuously revolve around us. But close observation showed that some planets moved forward in the sky, stopped, moved backward, stopped, and moved forward again. This could only be explained by all planets, earth included, revolving around another central point in the solar system, which he correctly concluded was the sun.

When Copernicus presented his ideas, he suggested simply that the sun does not revolve around the earth — the earth revolves around the sun. As we know now, his proposal was correct.

But after 15 centuries of belief that the earth occupied the prominent position at the center of the universe, the church took offense. They heard Copernicus' suggestion as a religious statement when, in fact, it was simply a statement of a new scientific theory.

Over the years, the Copernican Principle has taken many forms, but according to Gonzalez and Richards[165], it can be boiled down to a simple statement. It says, "We're not special."

The Roman Catholic Church was not happy about Copernicus' new heliocentric model. They understood him to be saying, "We're not special." In fact, they considered Copernicus' ideas to be heretical and anti-scriptural. Copernicus, however, died soon after publishing his ideas, so he did not receive any actual trouble from the church. But Galileo did.

---

[165] Gonzalez, *op.cit.*, p.286.

Galileo

As the result of studies using his newly improved telescopes, the Italian astronomer Galileo Galilei came to agree that Copernicus' heliocentric model was correct. Galileo was then labeled a heretic by the Roman Catholic church for his public teachings in support of Copernicus and his ideas.

Why did the Roman Catholic church adopt the geocentric theory into its doctrine in the first place? We may never know. According to the Bible, however, the earth and mankind **are** special and they **remain** special, regardless of how earth travels in its orbit.

Today's Understanding

The **Copernican Principle**, even today, is understood to say that the earth and mankind don't occupy prominent positions in the universe. That principle also gave rise to the broader **Mediocrity Principle**, which says that there is nothing special about mankind or the earth.

Why is this important? With the teachings of *uniformitarianism* and the theories of *Darwinian evolution* proposed in the 19th Century (see Chapter 1), which have absolutely nothing to do with God, today's society has mostly been educated consistent with these principles. Most people have been taught that there is nothing special about the earth or mankind. Since the two theories label the Bible's explanation of creation as a fairy tale, their proponents are perfectly happy to continue to teach both the Copernican and the Mediocrity Principles.

The Bible's View

Mankind **is** special, however, according to the Bible. Mankind always has been and always will be special. So, too, the earth!

Man is special because Adam was created in the **image** and **likeness** of God. **"And God said, 'Let Us make man in Our image, after Our likeness: and let them have dominion over the fish of the sea, and over the fowl of the air, and over the cattle, and over all the earth, and over every creeping thing that creepeth upon the earth.'"**[166]

---

[166] Genesis 1:26

Note also in this passage that man was given **dominion** over all living creatures. God placed Adam in the garden and gave him a very special assignment to fulfill. Mankind is very special to God and we can find examples of that all throughout the Bible.

The earth is special because it was created specifically to be **inhabited** by man. "**For thus saith the LORD that created the heavens; God Himself that formed the earth and made it; He hath established it, He created it not in vain, He formed it to be inhabited: I am the LORD; and there is none else.**"[167]

The earth was created especially for man and beast to occupy with man representing God as he took dominion over the creation. "**But now we see not yet all things put under him**."[168] The fact that mankind has yet to take proper dominion, does not affect the specialness of the positions of mankind and the earth in God's creation.

The Bible has no disagreement with the original heliocentric theory suggested by Copernicus. The earth and planets of our solar system really do revolve around our sun. But the Bible directly contradicts society's current understandings of the Copernican and Mediocrity Principles.

We are special. The earth is special.

---

[167] Isaiah 45:18

[168] Hebrews 2:8

6

Intelligent Design

Were the earth, mankind, all life, and the whole universe intelligently designed? ... or did they just evolve out of nothing? Many today are attempting to prove the need for intelligence in the creation. They are proponents of the Intelligent Design (ID) community.

Gonzalez and Richards provided us with a description of the current view of ID by the scientific community. "The 'official' view now among scientists and academics is that the notion of intelligent design is either unscientific or at least superfluous to the practice of all natural science."[169]

As carefully as those pursuing Intelligent Design stay away from religious arguments, the courts have recently taken a different view. In the case Kitzmiller v. Dover Area School District[170], the verdict came down in favor of the plaintiffs because "intelligent design is a form of creationism."[171] Teaching ID is therefore unconstitutional in public schools in that Pennsylvania district — and probably in the broader region as well.

ID proponents insist that belief in ID does not require faith in an all-powerful being. They insist they are not making religious arguments. They are just trying to demonstrate that creation could **not** have occurred mindlessly — intelligence was needed to create the world. The universe, the earth, and all living beings in it were designed — somehow. How?? If intelligence was

---

[169] Gonzalez and Richards, *op.cit*, p. 295.

[170] Kitzmiller, *op.cit.*

[171] Wikipedia, "Kitzmiller v. Dover Area School District."

needed, whose intelligence was used? That's anybody's guess! Any intelligence but God's could successfully answer that question.

The element of design in evidence in the created world, however, runs contrary to the other theories of creation like uniformitarianism and Darwinian evolution. They require total randomness — no intelligence is allowed.

If one does happen to believe that the God of the Bible is the Creator, that takes nothing away from the ID arguments. It just supplies the name and identity of the intelligent designer.

Obviously, the court in Pennsylvania didn't buy that argument since they think ID is a form of creationism. Regardless of that court's conclusion, however, many scientists are continuing in their attempts to prove ID, staying as far as possible away from even the most minute hint that the source of the intelligence is the God of the Bible. At any such suggestion, their arguments will have crossed into religious territory, and religious arguments are not science. End of discussion.

From where did the intelligence come? They don't say. Possible sources, I guess, could be a superior race of alien beings, outworlders, ETs, Vulcans, or any number of other crazy ideas. Any possible source is apparently permissible — except, of course, for the true Creator God.

A Religious Answer Is Not Automatically Incorrect

If God actually did create the heavens and the earth, as the Bible says He did, **labeling it a religious explanation, doesn't make it wrong**. It just means that it is an answer from the realm of faith — not science.

In the eye of public opinion, there are big differences between religious and scientific arguments. But the public is confused on this point because evolution is also a **religious explanation**. Evolution is religion, even though everyone insists it is science.

By supporting evolution as the only true means of creation taught in the public schools, the public thinks they are standing up for science. But half of all evolutionary 'science' is bogus (the macroevolutionary half). As the big boys have said, they just need a little more time to design the **right** experiments and they can prove evolution. They have had decades to do this, but they still have not succeeded.

If the God of the Bible created, as the Bible says He did, nothing would change in the book The Privileged Planet[172]. Gonzalez and Richards didn't disclose the source of the intelligent design in their book. They wrote the following, by way of explanation: "We must distinguish between an argument for design and an argument for the existence of God. While a successful argument for the design of the cosmos provides support for belief in the existence of God, it doesn't prove that the God of traditional belief exists. The most it establishes is that there is a designer sufficient to design the universe as we see it."[173]

Evolution??

Evolution, on the other hand, requires no intelligence. It doesn't even require a designer. Chemicals randomly came together to form amino acids, which randomly came together to form proteins, which randomly came together to form living cells, which randomly mutated into more complex, higher life forms. Ultimately, as you proceed up the chain of life forms, the lesser life forms randomly evolved into mankind. Every step was random. Intelligence and design were absent.

As mentioned, the facts to support Darwin's theory of evolution have never been proven, identified in nature, nor demonstrated. In fact, most arguments defending evolution compare apples and oranges. As we discussed earlier, evolutionists demonstrate the kind of evolutionary changes **that do exist** (known as microevolutionary changes), but they label it only ***evolution***. Then, they infer that since they have demonstrated that ***evolution*** exists, they have made their case.

But to prove evolutionary changes that can cause lower life forms to evolve into higher life forms, they have to demonstrate the other form of evolutionary changes — macroevolutionary changes. Such changes have yet to be demonstrated, proven to exist, or even identified in the fossil record. Evolutionists can draw wonderful pictures, make fabulous fictional videos, give greatly detailed explanations, make splendid, fully illustrated textbooks, etc.,

---

[172] Gonzalez and Richards, *loc.cit.*

[173] *ibid.,* p.330.

but to date, they have not been able to demonstrate **one** example of an actual macroevolutionary change.

The biochemist Michael Behe defined micro- and macro- evolution. "Roughly speaking, *microevolution* describes changes that can be made in one or a few small jumps, whereas *macroevolution* describes changes that appear to require large jumps."[174] He then added, "But it is at the level of macroevolution — of large jumps — that the theory evokes skepticism. Many people have followed Darwin in proposing that huge changes can be broken down into plausible, small steps over great periods of time. Persuasive evidence to support that position, however, has not been forthcoming."[175]

The scientific establishment and society continue to support evolution as "factual science." Most Americans, who have been taught evolution as fact, appear vehemently opposed to any suggestions that they have been taught erroneous information in school. High school and college biology classes and almost all other science classes teach evolution. It is difficult to make any headway when all of society believes that evolution is good, evolution is science, and science is fact.

The exceptions to this are students who have been home schooled, or who went to Christian schools, colleges, or universities. But they are in the minority.

For these reasons, any suggestion that God might have designed and created the heavens and the earth is treated by society as baloney. **Vast negative public opinion, however, cannot negate the truth.** It can suppress it for years and years, but it can't prove it wrong.

Since no one was there to see creation happen, **all creation theories come from the realm of religion. The truth, therefore, will come from the realm of religion.** On which particular system of belief will you choose to put your faith? On God? On random chance?

To the author, as an engineer, a Christian, and a teacher of biology, **the real baloney is** the idea **that the vast complexities of life are all the result of random chance and are completely devoid of design or intelligence**. The complexities of the human body and the human mind, of animal, fish, and

---

[174] Behe, *op.cit.*, p.14.

[175] *ibid.*, p.15.

plant life, and of the ordering of the earth and its many functions (known to many as characteristics of Mother Nature), all fall into this category.

Today's best and brightest (scientists, engineers, doctors) have not been able to design or build anything that comes anywhere close to approaching the complexity or the elegance of the human body. We don't even fully **understand** the complex, but elegant workings of the human body or these other examples. Yet we are supposed to believe (according to our best evolutionary "science") that the human body, life in all its forms, and Mother Nature evolved without the aid of intelligence. I don't think so!

The Bible's explanation is much more believable than that!

The Earth Was Designed for Exploration and Discovery

The main conclusion of the book, The Privileged Planet[176], is "The universe, whatever else it is, is designed for discovery." The key word in that conclusion is ***designed.*** Gonzalez' and Richards' whole book makes the case that our habitable planet — any habitable planet — was designed with everything it needs for space exploration and interstellar discovery.

Their attempts to calculate the probability of finding a planet with earth-like, habitable conditions yielded an incredibly small value — near zero.

To do this, they listed all of the conditions needed by a planet to support life. Such a planet must orbit the right kind of star, be at the right distance from the star, have the right magnetic field, have the right diameter, have the right tilt to its rotational axis, have a reasonably large moon, have the right atmosphere, reside in a system with giant planets in the outer orbits, etc. To each of these factors, they assigned a probability from 0.0 to 1.0 (0% to 100%). The probability that all such conditions are present in a single planet is simply the product of all the individual probability factors.

When their list includes 20 such factors, which they suggested is the typical number of factors required to define a habitable planet, the estimated probability of finding a habitable planet is on the order of 1 in 10^{15} (or 10^{-15}). They estimate there are about 100 billion stars (and solar systems) in the galaxy. (100 billion = 100,000,000,000 = 10^{11}) The product of these two numbers produces a very small number — near zero. This calculation says that

---

[176] Gonzalez, *op.cit.*, p. 311.

out of 10^{11} stars in our galaxy, 10^{-4} (or 0.0001) of them will have all the right conditions to support a habitable planet and complex life.

Their original assumption was to assign each of the probabilitiy factors the value of 10%, that is, 0.1. Using that number, the product predicts only 0.000 000 001 of the 10^{11} stars will have all of the right conditions. If five more conditions are required to produce a habitable planet, the probability of finding one with all the right conditions decreases further — to the order of 10^{-20} to 10^{-25}. With the number of stars remaining the same, the new product shows that the probability of finding a habitable planet becomes 10^{-9} to 10^{-14}. That is, 1 in 10^9 to 1 in 10^{14}.

If you believe in coincidences (and you must to believe in evolution), the earth and its abundance of complex life forms are all here due to a huge set of coincidences.

From a probability point of view, too many coincidences are required for the earth to be able to support complex life for it to be a totally random occurrence. You have a better chance of winning the lottery than a planet's chance that it will support life. Each condition required to support a habitable planet needs not only to be present, but it needs to be present within a very narrow range of acceptable values. If any one required condition has a value outside the acceptable range, the habitable planet could not exist.

Gonzalez' and Richards' final conclusion[177,178] was that the Earth "is very rare in the galaxy." The Bible agrees.

SETI

The Search for Extra Terrestrial Intelligence (SETI) represents mankind's attempts to find and identify signs of life in the universe. Carl Sagan estimated "that there might be one million civilizations in the Milky Way."[179] Lots of money was spent during the last 50 years on the SETI program to find

---

[177] Gonzalez and Richards, The Privileged Planet, DVD.

[178] Gonzalez, Guillermo, Preview Clip for The Privileged Planet DVD, "We're rare in the galaxy.", www.illustramedia.com/tppinfo.htm

[179] Gonzalez, *op.cit.*, p. 280.

and identify those intelligences. To date, they haven't found any. If Gonzalez and Richards are correct, the SETI project will not be very successful.

Are We Alone in the Universe According to the Bible?

The possibility of an unsuccessful SETI project brings this question to mind. The Bible doesn't address it directly. But the Bible clearly indicates that both the earth and mankind are special — contrary to the popular Copernican Principle.

If it doesn't address this question directly, what implications can we derive from the Bible. It tends to support the conclusion that "Yes," we are alone in the universe.

Why? God created the heavens and the earth[180] and He created the earth to be inhabited.[181] He placed mankind on the earth as His representative to have dominion over His creation.[182] Satan (a.k.a. Lucifer), God's most beautiful angel,[183] rebelled and was cast out of heaven — to fall to earth.[184] God's Son Jesus then came to earth to right the mess Satan, with Adam's cooperation, had caused. Jesus' solution was to exercise dominion over the earth and to solve man's sin problem. He did this "**once.**"[185] The verses that support these statements follow:

God created the earth for habitation by mankind. "**For thus saith the LORD that created the heavens; God Himself that formed the earth and made it; He hath established it, He created it not in vain, He formed it to be inhabited: I am the LORD; and there is none else.**"[186]

---

[180] Genesis 1:1

[181] Isaiah 45:18

[182] Genesis 1:26

[183] Ezekiel 28.15

[184] Revelation 12:7-10

[185] Hebrews 10:10, 12, 14

[186] Isaiah 45:18

God gave man dominion over all living creatures on the earth. "**And God said, 'Let us make man in our image, after our likeness: and let them have dominion over the fish of the sea, and over the fowl of the air, and over the cattle, and over all the earth, and over every creeping thing that creepeth upon the earth.'**"[181]

The devil, a.k.a. Satan or Lucifer, the highest created being, messed this all up. "**Thou wast perfect in thy ways from the day that thou wast created, till iniquity was found in thee.**"[182]

And when push came to shove, Satan was cast out of heaven from whence he fell to the earth. "**And there was war in heaven: Michael and his angels fought against the dragon; and the dragon fought and his angels, and prevailed not; neither was their place found any more in heaven. And the great dragon was cast out, that old serpent, called the Devil, and Satan, which deceiveth the whole world: he was cast out into the earth, and his angels were cast out with him. And I heard a loud voice saying in heaven, 'Now is come salvation, and strength, and the kingdom of our God, and the power of His Christ: for the accuser of our brethren is cast down, which accused them before our God day and night.'**"[187]

Jesus came to earth to right the wrongs of Satan, Adam, and Eve and to fulfill the mandate given to mankind. He successfully accomplished all this — but He did it only once! "**By the which will we are sanctified through the offering of the body of Jesus Christ once for all. And every priest standeth daily ministering and offering oftentimes the same sacrifices, which can never take away sins: But this man, after He had offered one sacrifice for sins for ever, sat down on the right hand of God; From henceforth expecting till His enemies be made His footstool. For by one offering He hath perfected for ever them that are sanctified.**"[188]

Summary

There is no suggestion anywhere in the Bible that God will place life on any other planet in the universe and try again. There is only one God, and He created earth using His intelligence, wisdom, and knowledge. Of all the

---

[187] Revelation 12:7-10

[188] Hebrews 10:10-14

possible planets in the universe, God created earth for mankind's habitation, and then He placed mankind here. Mankind fell, and God's Son came to fix the problem. There is only one Satan and he fell here to earth. There is only one Son of God, and He came here to earth to live among His people — and He promised to come to earth again.

That "God created the earth to be inhabited," answers how all the variables required to define a habitable planet just happen to be right for the earth. All of those phenomena were precisely adjusted by God.

The phenomena all have optimum values! The universe was intelligently designed — by God. Mankind is unique! The earth is unique! Our existence is very special in the universe! And this certainly is a privileged planet!

7

Computer Models

Now we need to look at computer models because they are at the center of the global climate change debate. I am very familiar with computer models because I have written some. I assume the average person, however, knows very little about them.

In this chapter, I am going to use my own models, and problems I had to solve, to explain modeling and demonstrate typical problems faced by all computer modelers. I will do my best to make it simple and understandable.

What Are Computer Models?

We see wonderfully capable androids and advanced computers all the time in science fiction books, on TV, and in movies. Sometimes we deal with advanced computers in our daily lives, but what kinds of calculations and predictions come out of *computer models*? What exactly are computer models?

21st Century citizens are spoiled because the average computer they carry around or use on their desk is an **advanced** computer compared to the computers we had 30-40 years ago. Yes — we had computers in the 1970s but the advanced computers of that day were huge. Some were equivalent in size to eight large refrigerators. Some were even larger. But over time, their sizes decreased as computing power increased.

All computers, even back then, used programming languages similar to those available today. Today's computers are much easier to use and program; they are much smaller; and they are orders of magnitude faster; but the principles of both hardware and software remain the same.

Computer models are computer programs that use simplified mathematical algorithms to simulate the behavior of real systems. For example, the author used computer models (1) to simulate the dense packing of particles, and (2) to predict the viscous behaviors of particles suspended in carrier fluids. Generally, computer models are developed to predict how a system of interest will behave under a variety of controlled conditions. Input some starting conditions and see what outcomes the model predicts — all performed by a computer in a clean, controlled environment.

Examples

For example, the author's research team wanted to identify the size distribution of particles that would pack most densely. We could have performed many lab experiments, or we could have developed a computer model. Actually, we did both. The computer models' results showed us which lab experiments to perform, and the lab experiments verified the computer models' results.

Our team also wanted to know what kind of viscosities we could expect from highly concentrated suspensions of powders. Specifically, our computer models helped us identify the major phenomena that controlled viscosities in dense suspensions flowing at high rates. Once again, the models predicted and the lab experiments verified the results.

Global Climate Change Models

In the study of global climate change, researchers are developing computer models to simulate global weather systems. They have their work cut out for them because, as you know, it is difficult to accurately predict the weather just a few hours from now. The climate change models being developed use recorded weather data (temperatures, pressures, relative humidities, etc.) to predict future climate behaviors. With their models, they wish to explore long time frames — that is, they want to predict climates decades into the future.

Possible Problems with Computer Models

The Math Is Inaccurate

Whenever possible, the simplest equations are used to simulate important variables. Under tightly controlled conditions, linear equations may be perfectly sufficient to simulate the behavior of more complex functions. When a phenomenon is known to have a non-linear behavior, but the range of interest of the independent variable is extremely limited, a simple linear equation still may adequately simulate the phenomenon's behavior.

Let's consider a viscosity example to demonstrate this problem. Viscosity defines how easily flow occurs in fluids. Molasses has a high, constant viscosity. Water has a low, constant viscosity. When powders are mixed into water, for example clay plus water, *suspensions* are formed. Unlike the viscosities of simple fluids, suspension viscosities are **not** constant. Initially, the viscosities of suspensions increase linearly as particles are added to a fluid. In those cases, suspension viscosities can be calculated using a linear function of the viscosity of the water and the concentration of particles added. As more and more particles are added, however, viscosities deviate from the linear behavior and the simple calculations no longer suffice.

As long as only a few particles are used, the linear function is adequate; it's easy to calculate; and it takes little time for computers to crunch the numbers. But as more and more particles are added (such as in pastes), viscosity calculations become much more complex because quite a few other physical phenomena contribute to suspension viscosities. When many powder particles and only little fluid are mixed to form pastes, collisions between particles dominate the control of suspension viscosities. Particle **collisions** MUST BE taken into account in such models.

As long as only a few particles are present in a fluid, the simple linear function to calculate suspension viscosity is perfectly adequate. A general use model intended to simulate the whole wide range of suspensions and their viscosity behaviors must use much more complex algorithms to take particle collisions and other controlling phenomena into account.

Good computer models should take into account all appropriate variables and every phenomenon that contributes to the behavior to be simulated. Defining a mathematical algorithm that can simulate each and every behavior can be difficult. The simplest algorithm that performs the necessary calculation is always the best. Sometimes, complex algorithms take too much time to calculate and simplifications **must** be made.

Simplifications always limit the broad use of computer models because they are almost always less accurate mathematically than more complex algorithms. This is something every modeler must take into account. It is a trade-off. Less precision and accuracy usually trade for faster computing times.

Some Phenomena Are Not Easily Modeled

Taking particle collisions into account in the suspension viscosity model presented lots of problems. Particle collisions are simply not easy to model. They require complex algorithms and long calculation times. Part of the problem was that several physical properties each affected the frequency and nature of the collisions. Some conditions made the particles behave like billiard balls. Some conditions made the particles behave like rubber balls. Some conditions made the particles behave like sticky, silly putty. The size distribution of particles present affected viscosities as well.

How does one take all of that into account? ... mathematically? Figuring out ways to take all of these properties into account and to develop adequate algorithms was extremely difficult.

Memory & Run-Times Are Limited

Another major factor in computer models always has to do with the size and speed of available computers. Our first lab computer thirty years ago could handle only very small computer programs. PCs have grown over the years. Today's size limits are enormous. Thirty years ago, it was easy to write a program that was too large for the computer. Today, it is almost impossible. Thirty years ago, the computers' speeds were extremely slow. Today, they fly!

It was possible to find larger, faster computers in those days, but payments to the computing centers were required based on the length of computing times. And that forced decisions that were trade-offs: How much could we afford? ... how long would it take to perform the calculation? ... how big could the program and memory requirements be?

One computer model closely simulated a microscopic analysis. Each iteration performed the tasks using random numbers to mimic an actual laboratory procedure. How many iterations should we run? 10,000 iterations of the simplest version of the program required several hours on a large, fast, IBM 370 Mainframe computer. 10,000 iterations of the more general (and more complex) model required overnight runs on the same computer. 100,000 iterations would have improved our results, but 5 days was too long to tie up the computer. A million iterations would have been great, but that would have required more than a month of computing time. We settled for 10,000 iterations.

Another model analyzed the packing of particles in a box. How many particles should we use? For random placement inside the box, each particle required three coordinates to define its position. If we wanted to simulate fibers, then we had to use three more values to define particle orientations. A model that would allow us to use real particle shapes would require much more memory than a model that packed spheres. Spheres required four words of computer memory for each particle we wanted to pack. Non-spherical particles required seven to ten (or more) words of memory for each particle. The number of particles we could attempt to pack was a function of the amount of computer memory available.

Then, to randomly place the particles inside the box, we had to determine whether they interfered with any other particles. That meant we had to check interference between each new particle and every particle already in the box. To pack the 100th particle, we had to check interferences with all 99 already packed. To pack the 1000th particle, we had to check it against all 999 already packed. If the particles interfered, we had to try another location. The more particles we tried to pack, the longer the model's run time would be. In other words, the first few particles would pack quickly while the last particles would be extremely difficult and time-consuming to pack.

The more general the program and the more types of particles we could simulate, the more useful it would be. But as the numbers of particles increased, memory requirements increased, the complexity of the computer program increased, and the time to run the models also increased. This especially applied to more realistic particle shapes (compared to spheres.) Memory requirements increased, the complexity of the algorithms increased, and run times increased.

How long would it take to write the general model and then, how long were we willing to wait to obtain good results? We settled on a fairly small number (about 100) of spheres (the simplest shape). That model could run easily on our small lab computer. I even programmed it to run 24/7 for several weeks when we would all be away from the lab. We obtained good results, but they were limited due to the trade-offs we made.

Mathematical Abilities Are Limited

Sometimes, models are limited by the mathematical ability of the modeler. Sometimes, models are limited by the programming skill of the modeler. Sometimes, complex phenomena can be ignored because those phenomena do not contribute under the conditions of interest, and by ignoring those phenomena, algorithms can be simplified. Sometimes, phenomena must be ignored until simulating algorithms can be designed. Attempts are always made to minimize limitations which can introduce errors.

Just because a modeler has a phenomenon that needs modeling does not mean that a model can be successfully created. Why? Each phenomenon must be well understood before a modeler can define simulating equations and design algorithms to include in a computer model. It is not possible to easily and accurately model any and all phenomena.

These limitations applied to all of the author's models, as well as to all of today's climate change models. We are still learning; our computer models are improving; but our overall abilities remain quite limited.

An Important Phenomenon Was Overlooked

Sometimes, phenomena that are important to a simulation are not included. Why? Because it is not obvious that the particular phenomenon will affect the results.

Of those who were developing models and studying the behaviors of suspensions, some apparently did not realize that particles collide with one another during flow; collisions are especially prevalent when lots of particles are suspended; and in crowded suspensions, those collisions dominate the control of viscosity behaviors.

In modeling runs with only two particles moving very slowly in lots of fluid, the particles will never collide. In such cases, collision algorithms can be, and were, absent from many computer models. When lots of fast-moving particles are flowing in close proximity, however, particle collisions must be taken into account.

Those resulting models which **did not** include collision phenomena could not demonstrate accurate viscosity behaviors because they had not included algorithms to simulate the important phenomena that control viscosities.

In one case, a research team was using such a model, developed without collision algorithms. They were advertising it, however, as a model that could simulate flows of practically anything. If it could flow, they said they could simulate it. But suspension models require collision algorithms.

That minor limitation, however, stopped no one from using that model on any and all systems. Their model always predicted flow results — results that ignored any and all collisions. Those results were bogus.

These examples demonstrate cases where models developed to simulate one type of behavior were used to simulate radically different types of behaviors. Of course, the models **always** produced answers — not necessarily correct answers, but answers. Since all appropriate phenomena were not included, the results were erroneous. But who knew? Who cared? They had a working model — and they applied it to practically everything.

For most of our models, we knew that more variables had to be identified, quantified, and included to properly simulate our systems. But we didn't know what those new variables should be. As the most critical phenomena are modeled and included first, remaining phenomena become less and less important, producing smaller, more minor changes in behavior. Complete accuracy becomes less easy to achieve when you are dealing with phenomena that produce diminishing returns. They **do** affect final behaviors, but they are relatively minor factors. Must they all be included? The correct answer is, "Yes." But in actual practice, many such phenomena are not included.

These types of problems are widespread — affecting all modelers and their developing models.

Which Problems Apply to Global Warming Models?

What has all of this to do with global warming models? Experience says that one or more of the above problems can easily be affecting climate change model results. It is highly unlikely that any of today's climate models are perfect. It is also highly unlikely that today's models are accurately simulating all appropriate climate phenomena. Have all pertinent phenomena been identified and included in the models? And if they have been identified, have they been properly simulated?

For example, Roy Spencer[189] has been saying for years that current climate models have not adequately accounted for the behavior of cloud cover. He repeated that concern in his newest book. Without proper algorithms to simulate cloud cover effects on weather conditions, today's predictive models can not produce accurate results. How does one simulate cloud cover? ... or the effects of cloud cover? I expect the simulation of cloud cover behavior is very difficult to effectively model.

How the models have been written and developed to simulate climate behaviors poses very important questions. Since we are being beaten over the heads with dire, catastrophic results from current models, everyone potentially affected by those predictions should be studying the details of those models which are very important to our futures!

Time Frames Are Important

When attempting to predict events a century or more from now, one must include all factors and cyclic phenomena that can occur within that time frame. El Niño and La Niña events cycle every few years according to Spencer[190]. It is fairly well known that solar activity follows an eleven year cycle.

Heating and cooling of our earth appears to follow a cycle with a period of 20-30 years. Why? Because, in the late 1970s, we were beginning to hear warnings of coming dire events — the alarmists were predicting that **global cooling** was about to cause a new ice age. And then — with a **SNAP!!** of the fingers — just like that, we were heating up and **global warming** concerns took over. How do each of these phenomena and their cycle lengths affect our climate? What are their natural oscillation periods? Do we even know for certain? It's unclear.

In his book, Roy Spencer suggested that the Pacific Decadal Oscillation (PDO) is important to any global climate predictions. It is a

---

[189] Spencer, Roy, The Great Global Warming Blunder, 2010, p.71.

[190] *ibid.*, p.15.

phenomenon over the North Pacific that affects weather patterns over the US. According to him[189], its cycle has an approximately 30 year period.

Have all such phenomena been adequately accounted for by the current models in the rush to judgement about global warming? Is there a comparable Atlantic Decadal Oscillation that affects Europe, or counterparts in the Southern Hemisphere?

Are there other phenomena that cycle over even longer periods that need to be identified and incorporated into today's predictive models? Cotton[191] commented that "deep ocean variability" affects climates and it has periods on the order of 100 years. It is doubtful that we have studied this phenomenon in sufficient detail to understand it very well. What data must be collected to properly understand and model this behavior?

If we have only been collecting good climate data by satellite for the last eight years, do we have enough good data to know whether or not we have taken all appropriate phenomena into account? It seems doubtful that we do.

Do All Phenomena Remain Constant?

When a model is designed to cover hundreds or even thousands of years, one must ask if each phenomenon has behaved and will behave similarly over the long time-frame? The assumption, of course, is that they do — they must!

But do they? Do all phenomena really behave the same way today as they did millions of years ago? Did they even occur millions of years ago? Do their behaviors vary slowly over long time intervals? ... or do they remain constant? Are the equations used to simulate these phenomena varied appropriately? ... or do they remain constant? The answers to these questions appear to be that we don't have sufficiently precise data for long enough periods of time to know definitively.

It seems less important to ask whether or not the modelers have used the right mathematical algorithms to simulate each phenomena, than it is to question whether or not these all-important phenomena have behaved the same way for millennia?

---

[191] Cotton, William R., "Is climate really predictable on 10-50 year time table?", Colorado State University, 20 Jul 2010, Powerpoint presentation.

To assume constancy, modelers must rely on the theory of *uniformity*. This theory suggests that all things today behave as they always have and always will. This theory forms the central premise of *uniformitarianism*, which was discussed earlier in this book. If *catastrophism* was actually involved in the formation of our earth, it is possible that some important phenomena have changed or will change over the projected durations covered by current climate change models. Has that been taken into account?

For example, if the earth's magnetic field is a factor in a phenomenon, is it constant or does it vary with time? Data shows that earth's magnetic field has varied with time — specifically, its orientation has changed (it has reversed). Do light waves coming to earth from the sun vary in intensity with time? Yes. If solar activity follows an approximate 11 year cycle, do the models accurately account for solar radiation variations? Are solar radiation variations accurately predictable? Maybe — maybe not. Similar questions must be asked about all phenomena when creating models that will predict results over such long time frames.

Where Are the Verifying Predictive Results?

Since we are constantly having dire *global warming* predictions shoved down our throats by anyone and everyone in the alarmist political arena and the mass media, where are the test results that predict today's (2010's) climate from a starting point at year 2002? Supposedly, we have been collecting good satellite data since 2002. Have model runs been performed to validate their predictions of today's climate using the good data? Have results of those runs been made available to everyone?

Every model I ever wrote had to be proven against solid laboratory data. When we said the viscosity of a suspension would behave a certain way, we had to prove it by measuring actual viscous behaviors of suspensions on the lab viscometer. When I said that analytical examinations of particle cross-sections under a microscope would produce certain results, I had to prove it experimentally.

Satellite data has shown that there has been no global warming over the most recent decade (from 2000 to 2010). Now we learn from the former director of the University of East Anglia's Climate Research Unit[192] (which is

---

192 Petre, *loc.cit.*

the research unit at the center of the global warming controversy) that no "statistically significant" global warming has happened since 1995. Can current models confirm this? ... or predict this?

The earth's temperatures over that time have fluctuated, but remained essentially constant. Why have we been hearing constantly about global warming during this last decade when none has actually been happening? The political elite and global warming alarmists can't even accurately report current temperature data. Most likely, they have been ignoring it. How, then, are we supposed to believe model results?

Since we have at least 8 years of good data, current models should accurately be able to predict 8 years out from the year 2002. Can they? If not, they need more work. If they can't accurately predict 8 years, why should we believe they will be accurate at 100 years? We shouldn't.

Where Are The Predicted Results Published?

If the predictions of climate change models are so dire and so critical to the survival of Earth, why aren't they published out in the open press where everyone can easily find them? Why are we always hearing about model results from unbelievable sources like the former VP or other politicians who are obviously attempting to further their political agendas? Rather, we should be seeing these reports published *en toto* in our daily newspapers. If they are so important, why not?

Unless we search appropriate technical publications, we will not see the actual results. Even if we find them, will they mean anything? When politicians and editorial page writers, representing groups who are not generally known for their technical literacy, are quoted in the daily news or on the evening news report as saying, "Of course this heat wave is the result of global warming!", shouldn't we be shown the actual results and details of the predictive models?

Over the last ten years, we frequently heard that the current heat wave is caused by global warming. Last winter's extreme cold was caused by global warming. The hurricane seasons since Katrina were all going to be especially violent due to global warming. The violent hurricanes themselves are caused by global warming. These all make great stories on the evening news. But we don't want to hear politically correct fluff! We want to hear someone who is believable — someone other than a politician, an environmentalist, or a news anchor. We want to hear explanations regarding how and why global warming caused each of these events. We should insist on this, especially in light of current news that says we haven't had any global warming over the last 15 years.

As mentioned, in the years following Hurricane Katrina, we were supposed to be hit by lots of violent hurricanes. Global warming, you know! Oh, it was going to be especially bad!! Since global warming has been worsening, new storms would be worse in the years following Katrina. The year after Katrina, none of this happened. Who goofed? The PC community was relatively quiet — and the newspapers and TVs were mostly silent as well.

Predictions like that still continue today. As long as they don't back off their predictions, one of these years, another large hurricane will hit the Southeast. Then we'll hear, "See — I told you so!!" They will ignore the several quiet years with all the erroneous predictions, to make this point.

The average citizen doesn't care about our ability to calculate viscous properties of suspensions, so those results remain only in the technical literature. But everyone is interested in global warming — especially when the current administration plans to use the models' results to change policies and wreak havoc on our capitalist economic system. For that reason, any pertinent model results should be broadly published and widely available in the daily international press.

When such results remain buried in the technical literature, citizens begin to wonder how truthful the harbingers of doom really are. Could it be that current results do not predict the politically correct, desired behaviors, so results remain hidden from view? Could it be that the alarmists want to stifle debate — especially when results are not to their liking?

Are Logical Explanations Forthcoming?

Several years ago, there were reports that Earth, Mars, other planets, and some of their moons were all experiencing increased surface temperatures and global warming.[193] The temperature increases on Earth and Mars were of approximately the same order. The initial suggestion by a Russian scientist was that the warming trend was due to the sun — which is a factor common to all.

---

[193] Than, Ker, "Sun Blamed for Warming of Earth and Other Worlds," LiveScience Staff Writer posted: 12 March 2007 07:27 am ET.

"No — that was not the case," we were told. Mars and other planets have been heating because of the sun. The Earth, however, was heating because of its socio-industrial complex.

That makes no sense. At first glance, one assumes that all planets should be heating due to the increased influx of solar radiation. The answer, "No!," was not accompanied by any good, logical explanations. It should not be expected to convince anyone — certainly not a skeptical public.

Why wasn't Mars, like earth, heating because of its socio-industrial complex? ... or why wasn't Earth, like Mars, heating because of increased solar radiation? The experts said that in Earth's case, it is well known that **the sun is not the primary cause** of our global warming. Really?? Why? Has the discussion ended before it actually started? Which came first? ... the chicken or the egg? Which came first? ... anthropogenic global warming or increased solar radiation?

This seems to happen all the time regarding global warming topics. We were told by politicians and the news media that the oceans are going to rise 26 feet in the coming years. Those reports are given with calls for urgency that we must do something quickly! But test results and technical data from the researchers are never easily available — and any outspoken skeptics draw nothing but ridicule.

We are expected to blindly believe the conclusions of politicians who have traditionally not received good marks for their veracity! The former VP is making lots of money off of his global climate change beliefs — and we are expected to take his word for it that mankind is causing the warming. The fact that he has a movie, a book, an Oscar, and a Nobel Peace Prize does nothing to advance our trust in him. Those four just suggest he is a good storyteller.

We are supposed to believe our news media, because they are supposed to be accurately reporting every important bit of news. In my own experience, whenever we were interviewed for a news story on our research, our activities and results were always misrepresented and erroneously reported. Sometimes, the reports included inaccurate but politically correct commentary to make the story more interesting.

Today's news media are known to ignore important stories. When they do "report" stories, the reports are frequently interspersed with commentary and opinion. The news is supposed to be factual. Opinions are usually printed under different headings — usually on editorial pages. Just ask General McChrystal, who was fired by the President as Commander in Afghanistan (23

June 2010) for outspoken remarks made in front of an embedded 'reporter,' whether or not the news media can be trusted.

Can A Computer Successfully Model Weather that is Controlled by God?

If it is true that God controls the weather, the wind, and the rain, as discussed in the Bible and in this book's Chapter 4, can current computer models accurately predict the weather or climate? The problem is that we have no way of knowing what God wants at any particular time, so the models will be lacking very important input data and effective algorithms. Who knows the mind of God? Didn't God ask Job that very question?

I have seen two cases in my industrial career where an engineering manager had to be considered an independent variable in the solution of daily engineering problems. Both managers were known to make major changes over the weekends. Sometimes, they imposed limits on testing variables for no apparent reason. Sometimes, they changed several variables at once — just when it appeared they were homing in on a good answer by fine-tuning a single variable. Often, their decisions changed production conditions.

When an independent mind is a variable in a predictive model, it is like shooting in the dark at a moving target. One never knows what will happen next.

In this case, we may know **exactly** how our weather systems work (we don't, but let's assume we do), but if God decides that it needs to rain over South Carolina today, He can make it happen. Even when all data suggest drought conditions are imminent, He can make it rain. He can make the wind blow. He can hold back the rain. He can do whatever He wants. How do we predict climate under those conditions?

If God is actually controlling the weather, as the Bible says He does, then computer models will **never** accurately predict future climate conditions. Why? The models will not be able to account for the mind of God!

Can A Computer Successfully Model Climate A Century Out?

This is another fundamental question. Even with our best efforts, can a computer successfully model weather at all? ... or decades from now? ... or centuries from now?

This is the question asked in a recent presentation by W.R. Cotton, Professor of Atmospheric Science at Colorado State University[194]: "Is climate really predictable on 10-50 year time table?

His conclusion reads, "Considering the stochastic external forcing parameters(e.g., volcanoes), uncertainties of solar variability forcing, and the tendency for strong model biases on time scales of 2-5 years let alone 10 to 50 years, I see no evidence that climate is predictable on these time-scales nor will it be for decades to come!"

Here we have a scientist in the right field, who questions whether or not we can model climate at all — while the alarmists have declared the debate to be "concluded."

The alarmists want us to put complete faith and trust in current computer models and here, an atmospheric scientist is questioning whether such predictions can be made at all.

It is an absolutely amazing world in which we live!!

Summary

Computer models use simplified mathematical algorithms to simulate real phenomena. Sometimes, the math is overly simplified and inaccurate. Sometimes, unknown but important phenomena have been excluded from models. Sometimes, we don't know how to simulate a particular phenomenon. Sometimes, accurate values of important variables are not available. Sometimes, validating tests which demonstrate a model's efficacy have not been reported — or are not readily available.

The author suggests that model results should not be believed, unless they come with verified test results that have accurately predicted the current climate when using the best available satellite data starting at 2002.

If we are expected to change our economy, and ruin our business environment with radical new taxes and income redistribution procedures to prevent the dire consequences predicted by current models, then we the people should INSIST that the modelers prove that their models work. If they cannot do that, we have no reason to believe any of their predictions for the 22nd Century.

---

194 Cotton, *loc.cit.*

Computer models are extremely complicated algorithms that can predict very precise results. But remember with computers: GIGO. Garbage In, Garbage Out. If a model's level of sophistication is not very great or if the input data is not very accurate, the model's results will also not be very good, accurate, or precise.

Global climate change phenomena are some of the most complex, difficult to understand, and difficult to model phenomena in existence today. The fact that we cannot accurately predict today's weather just means that current models need further improvements.

The most important idea to be taken from this chapter is the realization that if God controls the wind and the rain, it will be impossible to accurately predict climate. After all, who knows the mind of God?

If climatologists and meteorologists could predict today's weather accurately, that would be a great start. Then maybe, we could begin to have better confidence in their computer models. We could begin to believe that their predictions for next week's weather may be accurate. Whether or not they will ever be able to predict the climate hundreds of years into the future (even without God's controlling influences) is questionable.

By the way, today was supposed to be hot and sunny. (It's currently raining.)

Feedback Control Loops

Self-Regulating Phenomena

When one designs a control system for an industrial process, it is vital that good control loops are used. A good control loop utilizes a measurement sensor, a setpoint, a controller, and a control mechanism to affect the phenomena of interest. Good control loops are difficult to design — especially when the system to be controlled is complicated.

Some systems, however, are *self-regulating* and they don't need any external control circuitry. They consist of phenomena that regulate themselves naturally.

In an industrial setting, if a water valve over a tank is opened, with an open drain at the bottom, the tank will fill to the depth where the pressure (from the depth of water) forces the flow of water out the drain to equal the input flow. When flow of water through the input valve equals flow of water out the drain, the depth of water will hold steady at that level. The depth of water in the tank is self-regulating at that depth. A little more input flow will cause the level to rise a little bit; a little less input flow will cause the level to subside a little. Too much input flow will cause the tank to overflow, but even then, its depth will be regulated at the full depth of the tank. (It will make a mess, but it will be regulated.) Too little input flow will cause the tank to remain essentially empty as all water entering immediately flows out the drain.

A Well Designed Automatic Control Loop

It is highly unlikely that an engineer in an industrial setting would design a self-regulating system or use a self-regulating process. Such cases do not afford the engineering staff full control over the system. So even when a system is considered to be self-regulating (like the previous example), a good engineer would add a sensor, controller, setpoint, and actuator so he has complete control during processing. When designed properly, a good, automatic control loop allows the engineering staff full control over the process.

To change the conditions of the last example a little, consider a fluid flowing into a tank in which the level is supposed to remain steady at a certain depth as process requirements define how much liquid exits the tank. In this case, engineers will install a sensor to detect and transmit fluid depth to the controller. The desired depth (the setpoint) will be an input parameter to the controller. The controller will then open and close the input valve to maintain the proper fluid depth in the tank, regardless how fast the fluid exits the tank.

A control loop includes a sensor which measures the controlled variable (the depth of fluid); the setpoint value tells the controller how full the engineers want the tank to be; the controller does the thinking as it compares the actual depth to the setpoint depth; and finally, the controller opens or closes the input valve to maintain the desired fluid level in the tank. The output valve can open or close in response to downstream process needs, but when this control loop is designed and implemented properly, fluid depth in the tank will remain constant at the desired level.

Natural Systems

The earth's weather and climate systems are naturally self-regulating. The sun shines on the earth and heats the land, water, and all objects it encounters. The land, water, and all objects rise in temperature in response to the absorbed sun light. Exposed liquid water heats and at some point, it vaporizes and converts to water vapor. Water vapor rises in the atmosphere to form clouds, which block the sunlight from reaching the surface of the earth. The clouds reduce the localized heating, allow the land, water, and objects to cool, and slow the production of more water vapor.

When there is too much water vapor in the atmosphere, it condenses in the form of clouds, it rains, and it deposits liquid water back onto the earth.

Excess liquid water on earth runs into our lakes and rivers, and flows to the sea. Since a majority of the surface area of our planet is covered by water, the water in our oceans is heated by sunlight, some is vaporized, and the vapors return to the sky to form clouds. And so the cycle continues.

Natural systems like this are self-regulating. They are much more complex, however, than in this simple example. There are many more variables that affect the production of water vapor, clouds, precipitation, and run off than described above. There are more factors that influence the rate of water vaporization than just the amount of sunlight released by the sun and the amount of cloud cover.

Regarding climate, lots of complex phenomena with lots of influencing variables form the self-regulating cycles we know as weather and climate.

Note that if the earth is millions of years old, as we are led to believe by today's science, the self-regulating nature of the earth's climate system is clear. We have seen relatively constant climate conditions from year to year, season by season which demonstrates to us that the weather and climate systems work pretty well. And they have worked this way for thousands of years. We don't see big, man-made umbrellas that automatically open to shade us from the sun. We see clouds that shade us from the sun, but they are formed by natural phenomena and they participate in this complex, natural, self-regulated control system.

If catastrophist theories (as opposed to uniformitarian theories) are responsible for the shaping of our current world, then major cataclysmic events of the distant past helped to shape the earth into its current form. If that happened, the self-regulating nature of the earth's systems had much more extreme aberrations to work through — and they did so very well. A system that works with slow, smoothly changing weather conditions can appear to be a great control system, but it will really be tested when major, cataclysmic events occur. Our weather systems work well as they accommodate the unpredictable but frequent earthquakes and volcanic blasts. Our self-regulating weather and climate systems are very capable control systems. Our weather remains quite stable within livable temperature ranges.

Did these self-regulating systems develop by random actions in this earth? Uniformitarians would answer, "Yes." Intelligent designers and creationists would answer, "No." This author believes God designed them. He is a wonderful design engineer.

Stable Control Loops

The goal when engineers set up control loops is to cause a variable to achieve steady conditions that follow our desires. Consider a kitchen oven. We set it to 325°F and we expect the electronics to control the oven chamber so it heats to and maintains that temperature for the duration of baking. Kitchen ovens all have simple controllers built into them to produce steady temperatures at the desired values.

In the case of natural systems, relatively stable conditions are produced by our self-regulating weather and climate systems. We have temperature and weather variations from summer to winter and from year to year. And we have the occasional aberrations — earthquakes, floods, volcanoes, tornadoes, and hurricanes — but even following extreme aberrations, climate conditions remain quite stable. The occasional extreme storm upsets stability, but following a bad storm, conditions usually return back to normal fairly quickly. After a tornado has passed through and destroyed a neighborhood, more often than not, the news cameras brought to the scene to show the damage also show a bright and sunny day. The same is true after hurricanes pass by. Our self-regulating systems work very well!

Unstable Control Loops

An unstable control loop, on the other hand, is one in which the controlled variable behaves in a crazy, uncontrolled manner. A control system that cannot properly maintain the controlled variable at its desired value is a bad system. If the controlled variable fluctuates wildly, that system is out-of-control. Wild, random fluctuations are produced by unstable control loops.

Here's an example: When driving on ice in rear-wheel-drive cars, it is relatively easy to cause the car to fishtail. When corrections are applied properly to the steering wheel — that is, with the right amount of correction at the right time — fishtailing can be controlled and the car can be brought back under control. It's quite fun to play with fishtailing, especially on an empty parking lot during winter snows. When the corrections are made erratically, too extremely, too quickly, or too slowly, the car can spin out — that is, it can go totally out of control. It may be fun, but it indicates bad control of the car.

In industrial systems, out-of-control conditions are not fun, valuable, nor desired. There is no room for experimentation or playing around in industrial control loops. They need to be designed right and implemented well from the beginning.

Industrial control systems utilize a feature, known as the *gain*, to tell the controller how much correction to make for any given error. That is, the *gain* controls how much correction is made when the controlled variable's measurement deviates from the setpoint. When *gains* are adjusted improperly, industrial control loops can produce out-of-control conditions.

Most industrial controllers have lots of flexibility in the way they impart corrections. A good controller, known as a PID controller, allows big corrections (D) immediately at the first sign of an error, lesser corrections (P) that are proportional to the error, and slow corrections (I) as long as the error endures. Three gain adjustments, one for each of the functions, P, I, and D, allow complete flexibility of such control loops. When the gains are adjusted properly, a PID controller can return the controlled variable very quickly to its desired, setpoint value.

For example, to heat an industrial furnace to a particular temperature, gas burners are frequently used. An electronic controller handles the control function. The desired temperature during the firing cycle, called the *setpoint*, is entered into the controller. A sensor measures the temperature in the furnace and sends that value to the controller. The controller compares the setpoint value to the actual (measured) value, calculates the error, and corrects the flow to the burner, combining the gain setting with the magnitude of the error. It increases or decreases the amount of heat needed to bring the measured temperature back to the setpoint. When the control loop is designed and implemented properly, the temperature in the furnace will sit steadily at the setpoint throughout the whole firing cycle. If the setpoint changes with time throughout the heating cycle (that is, if it follows a heating curve), a good controller will cause the actual temperature to track with and match the setpoint at all times.

When the *gain* settings are adjusted improperly, the furnace temperature can vary widely, wildly, and uncontrollably away from the setpoint. Under out-of-control conditions, chamber temperatures can be too low to achieve proper heating conditions, or too high to produce proper heating. Cakes, bread, and roasts can be burnt in kitchen ovens with faulty controllers. Ceramic wares and furnace insulation can be melted and/or destroyed in kilns with faulty controllers.

Sometimes, burners and furnace temperatures cycle too quickly, producing wild temperature swings (from extremely high temperatures to extremely low temperatures) during the firing cycle. Sometimes cycling like this is too wild to be brought under control. In such cases, the controller will not be able to reduce the temperature swings to achieve steady temperatures. All of these possible behaviors are the result of the gain settings on the controller.

As mentioned by Spencer in his book[195], the sign (+ or -) of the gain can also be wrong. For example, when the furnace temperature is too high, the controller should decrease fuel input to the burners. If the controller's gain setting has the wrong sign, when the furnace temperature is too high, the controller will increase the fuel input to the furnace — making the temperature hotter, not cooler.

Having the right sign doesn't guarantee stability of control, but having the wrong sign usually guarantees out-of-control systems.

Earth's weather and climate systems don't have this problem. They are self-regulating systems. We can't make any adjustments to help them along. But they don't run out-of-control. And they haven't been out-of-control as best we can tell.

Yet when we design automatic control systems, out-of-control conditions are always possible.

Computer Models

Even though natural systems like the earth's climate and weather are self-regulating, all of the phenomena that contribute to the various cycles must be incorporated into and simulated by computer models. To model the various natural cycles that participate to produce our weather or our climate, it is necessary to determine how all of the various phenomena interact with each other. How is this done? We must identify the precise *gain* settings to be used in the model as if we were designing an external, automatic control loop. The natural process doesn't need our gain adjustments, but **the computer model**, being developed to simulate the natural system, **does**. Entering the wrong gain values into computer models will produce problems, just as if we entered the wrong gain values into a real, automatic control system.

---

195 Spencer, *ibid.*

We can't alter any gains that govern the natural control systems on this planet, but to simulate them in computer models, we need to determine the precise values of the gains used in the natural control systems. We can't develop models without clear understandings of the working of the real systems.

Spencer[196] suggested that the computer models that are predicting dire climate consequences are using improper *feedback* settings. His term *feedback* is equivalent to the term I have referred to as *gain*. Feedback and gain both tell how much correction, and what kind of correction, needs to follow a change in conditions. When incorrectly set, both can easily allow controlled variables to run out-of-control.

What exactly does Spencer think is wrong in the current batch of climate models? He believes that the sum of the feedbacks in most current models are positive, which is the sign (+ or -) of feedback that allows out-of-control fluctuations to occur. Negative feedbacks allow systems to settle back towards normal conditions.

His argument suggests that the interactions between one or more of the variables in the climate models are incorrect. How can one sum the results of several variables to produce the oppositely signed feedback, if all feedbacks are negative to start with? Something is wrong.

Evidence from recorded history, which says our weather and climate are under pretty good control, suggests something is wrong with the models.

For instance in the furnace example, when the temperature in the chamber is too high, you want to decrease the input so the temperature comes down. That requires a negative *gain* (or *feedback*.) If the *gain* is set positive, (which is the wrong sign) and the temperature is too high, more heat will be added and the temperature will actually rise more. That will lead to a greater error, more heat added, and an out-of-control system.

When including many variables to simulate complex systems such as climate, incorrectly assessing the functioning of variables (that is, incorrectly assessing individual *gains*) and the **interrelationships** between variables (incorrectly assessing how one variable change affects another variable) can lead to controlled variables that don't (or can't) easily approach steady state conditions. This can produce unrealistic predictions.

---

[196] Spencer, *ibid.*, p. 54 *ff.*

Complex interactions of variables define systems that are difficult to fit with appropriate automatic control systems. They are equally difficult to accurately model. Any one particular input variable can affect many other variables, and the whole collective effect defines the complete working of the system. Any one *gain* setting may be approximately correct, but the collective influence of all *gain* settings will control the whole system.

If what Spencer has suggested is true, at least one *feedback* value, and maybe several of them, may be completely off base. Weather and climate systems are extremely complex, but the natural system works well. Because they are complex systems, it will be very difficult for modelers to set and balance all *feedbacks* properly, and prove the models' reliability.

The earth has demonstrated, over and over again throughout its history, its ability to handle wildly varying conditions and events and still produce relatively steady weather conditions. Wild storms and volcanos happen routinely, and our weather is still relatively stable. Earth's natural control system appears to be a very robust system which can handle just about anything.

Are today's climate models equally robust?

Summary

When modeling natural phenomena that are self-regulating, and which historically have produced relatively steady temperatures, sea levels, winter weather, summer weather, etc., current models should also be expected to predict well-controlled results. Do they?

All climate change models should demonstrate their capabilities using the last eight years of excellent satellite data.

It is unrealistic that models should use *feedback* values that allow uncontrolled variable swings for those variables (temperature, for example) that are known in the real world to be relatively stable. How each variable is controlled and how each interrelates with all other variables should be broadly examined and explained by modelers.

Each individual algorithm should be capable of demonstrating proper, steady simulation of its variable. And then, the collective algorithm should be capable of demonstrating how the whole, integrated system works. The results of such demonstrations should be widely published and available to all interested parties before they are used to predict dire consequences centuries from now.

The public should not need to take the word of non-technical cheer leaders (read: the political elite) that current computer models are accurate predictors of 22nd Century climate. If we are going to be beaten over the heads in the daily press with global warming hysteria, then those same media outlets should make the actual models and results available for all to view. If they have nothing to hide, they should do this. Then, we can all know how well their models simulate earth's well-running, natural, self-regulated weather and climate systems.

Current data shows no global warming over the last 15 years. **Can** the climate models show that as well? **Do** the climate models show that?

9

Important Natural Cycles

There are quite a few natural cycles that need to be considered when discussing global climate systems. The Water cycle is a primary example of a natural cycle on earth. Liquid water absorbs heat from the sun, vaporizes, rises into the atmosphere where it forms clouds, and eventually produces precipitation (e.g., rain — liquid water) which falls back to earth to start the cycle again.

Another natural cycle is the Dust to Dust cycle from the Bible: God said to Adam, **"In the sweat of thy face shalt thou eat bread, till thou return unto the ground; for out of it wast thou taken: for dust thou art, and unto dust shalt thou return."** (Genesis 3:19) All living creatures on earth have this in common: for dust thou art, and unto dust shalt thou return. Our human bodies average about 57% water[197]. After a body has been cremated, a small urn full of ashes is all that remains. The non-aqueous components in our body come from the dust of the earth. After death, the organic materials in our bodies decay to release the ashes (minerals) back to the earth.

There is a Carbon/Oxygen cycle. Humans and animals breathe in oxygen and exhale carbon dioxide. Plant life uses carbon dioxide during photosynthesis for growth and gives off oxygen as a by-product. As long as populations of human and animal life remain in relative balance with the earth's plant life, sufficient oxygen and carbon dioxide for each will remain in our atmosphere.

A variation on the water cycle occurs when moisture evaporates from the oceans of the world to produce water vapor and clouds. These produce

---

[197] Wikipedia, "Body Water."

precipitation of all forms (rain, sleet, hail, snow) over the earth. Excess ground water runs off to form streams and rivers which allow the water to flow back into the ocean.

Alternatively, when the sun shines on the earth and its oceans, it heats the surface. When the earth and its oceans are too hot, clouds that form from vaporized water can block the sun light from reaching the earth's surface. Cloudy conditions cool the surface back towards original values. Eventually, water vapor condenses, precipitation forms, and the sky clears again.

Quite a few such cycles and variations are interrelated. Water vaporizes to form water vapor which is a gas. Water vapor condenses to form clouds. Clouds hold rain, snow, and hail. Clouds block sunlight, cooling the earth. Rain and other precipitation waters the earth, fills streams and rivers, provides drink for the living, sinks into the ground to replenish the water table, flows to sea, etc.

Some interactions can be rather complex. But all such phenomena come from natural, self-regulating processes.

One belief is that the Creator God (see Chapter 2) set up all of these cycles when He designed the earth. Another belief is that evolution and uniformity randomly caused the development and utilization of all these cycles. Let's give thought to the meanings of some special cycles.

The Oxygen/Carbon Dioxide Balance

Oxygen and carbon dioxide form an important cycle — possibly **the** most important cycle on the planet. All life participates in this cycle. We'll start with some background information and then work our way into a discussion of the actual cycle.

There are many oxide materials that make up the earth's crust. Deposits of these raw materials are scattered around the earth. Most oxide minerals are metal oxides formed due to the high concentration of oxygen in our atmosphere. Metal oxides and other more complex oxides fall into the category known as **inorganic** materials.

Human bodies, other living creatures, and vegetation are members of the other category of materials known as **organic** materials. Most organic materials, many of which are known as **hydrocarbons**, are based on carbon chains and structures with attached ions such as hydrogen, oxygen, nitrogen, and sulfur. The term hydrocarbon suggests that they are mostly formed from hydrogen (hydro-) and carbon (-carbon). Fossil fuels are relatively pure hydrocarbons, but complex organic materials form living tissues, foods, polymers, oils, plastics, pharmaceuticals, household chemicals, soaps, and many other common and useful products.

The best examples of organic materials are our fossil fuels: natural gas, oil, and coal. All three are considered to be **renewable resources**. It just takes many years longer to produce crude oil and coal from decaying plants and animals than it does to produce natural gas.

Early stages of decay of organic materials produce natural gas. For this reason, many land fills are being fitted with vent pipes to collect the natural gases given off by decomposition products from compacted garbage.

Our enormous reserves of crude oils and coals come from vast deposits of decaying vegetation and animal life. If we truly have as much oil and coal deposits as our identified reserves suggest, one might wonder where the vast supplies of precursor materials (the compacted, decaying plant and animal life) came from? Enormous quantities of vegetation must have grown over the face of the earth at some point in earth's history, and then been buried and compacted, to form these resources.

Now let's look at the natural cycle: What happens when we burn fossil fuels? The global warming alarmists want us to believe we are ruining the atmosphere as we burn fossil fuels. But complete combustion of hydrocarbon fuels produces water vapor and carbon dioxide, both of which are needed to support life on this planet. The growth of plant life relies on ground water and carbon dioxide from the air. Through photosynthesis, plants take in carbon dioxide, convert the carbon into solid vegetation, and give back oxygen to the atmosphere. Growth of human and animal life relies on a continuous supply of oxygen in the atmosphere, taking in oxygen and returning carbon dioxide to the atmosphere as we breathe.

The complete carbon dioxide/oxygen cycle therefore includes the following: (1) humans and animals eat fruits and vegetables, which are converted internally to help our bodies grow and to produce energy; (2) the oxygen we breathe reacts with organic nutrients in our systems to produce energy for life and growth, water, and carbon dioxide; (3) we exhale carbon dioxide and excrete excess water; (4) plant life around us utilizes the gaseous carbon dioxide and liquid water for their growth and energy production; (5) the plants return gaseous oxygen to the atmosphere for humans and animal life to breathe; and (6) the plants provide their solid tissues, grains, fruits, and vegetables for food for human and animal consumption, for many other useful products such as rope, timber, etc., and excess plant volume can decay to produce more natural gas, oil, and coal. And so the cycle continues.

Another Carbon Dioxide Cycle

Alternatively, atmospheric carbon dioxide can be dissolved by rain drops in the air and brought to the earth's surface. The slightly acidic water helps in the weathering of minerals as it flows into earth's streams and rivers, and back into the oceans. Once in the water, dissolved carbon dioxide then can be used by algae and other under-water plant life, through photosynthesis, to once again produce oxygen and water. In this case, dissolved oxygen then becomes available for fish and other living sea creatures. They use the oxygen and convert it back to carbon dioxide for this natural cycle to continue.

A Natural Balance

Considering that humans and animals rely on oxygen in the atmosphere, and vegetation relies on carbon dioxide in the atmosphere, there is a natural balance in our world between these two major groups.

Studies have shown that plants thrive when carbon dioxide concentrations in the atmosphere are slightly higher than current levels, especially at slightly warmer temperatures. We also know that today's world population is as large as it has ever been. It makes sense then that as more people are alive today than ever before, plant life would benefit by elevated concentrations of carbon dioxide. In such an atmosphere, worldwide fruits, grains, and vegetables would thrive so a sufficient quantity of food can continue to grow and feed the world's population. This balance is critical to life on earth, and it is a perfectly logical balance when one thinks about and understands it.

If our fossil fuel reserves all came from decayed vegetation, as we think they do, one can assume that all fossil fuels came from once airborne carbon (carbon dioxide) and liquid water. The currently elevated concentration of carbon dioxide in the air, produced by human and animal life and supplemented by fossil fuel combustion, is providing the atmosphere with sufficient carbon dioxide to allow plant life to not only thrive but to once again produce the vast quantities of vegetation needed to renew the fossil fuels.

Combustion releases and returns the carbon in fossil fuels back into the atmosphere in the form of gaseous carbon dioxide. This completes its natural cycle. Vast quantities of vegetation grew in the distant past from atmospheric carbon dioxide. They were compacted and covered by the earth, where they decayed to form fossil fuels. Combustion of the oil and gas we collect and the coal we mine returns the carbon dioxide back to the atmosphere, where it is once again available to assist in the growth of vast quantities of vegetation.

The natural balance between atmospheric oxygen and carbon dioxide helps to maintain the balance between human and animal life and vegetation. Both categories are growing abundantly in today's atmosphere.

Why anyone would want to do something to adversely affect this natural cycle is beyond me! Yet that is exactly what the social elite want to do.

If society's intelligentsia want us to lower our carbon dioxide output, then I suggest that we should all comply only when we see all members of the intelligentsia skipping every other breath. This may be a cynical attitude, but so be it. It is such a fundamentally natural cycle, it appears to be sheer lunacy to mess with it.

This cycle is a well-designed cycle that insures the preservation of abundant life on this planet. There will always be people and animals, and they will always have a sufficient food supply. The cycle appears to be intelligently designed. In fact, one might ask how it just happens that plant life needs the waste products (carbon dioxide) of human and animal life? ... and how human and animal life just happen to need the waste products (oxygen) of plant life? ... and how it just happens that plant life is a staple food for human and animal life? ... and how human and animal life (through waste, death, and decay) is a staple food for plant life?

This cycle looks like the kind of well-engineered design that only an omniscient Creator God could produce. It seems almost impossible that such coincidences would be the result of the random chance of evolutionary theory. But that is exactly their answer — it's merely coincidence or chance.

Conclusions

Why would anyone want to alter this important natural oxygen/carbon dioxide cycle as a ruse to redistribute wealth from earth's rich to its poor? That is the perverted logic of mankind.

Maybe the God who created the heavens and the earth knows that to support as many people as we now have on earth, higher levels of carbon dioxide and slightly warmer temperatures are required to grow the necessary amounts of food, grain, and vegetables. Has anyone considered that? Yes, it requires faith in God, so it is automatically dismissed as a possible answer by many.

Not only are the high levels of carbon dioxide causing world problems, the highest levels of 'pollution' are supposedly coming from North America — from the US in particular. I suggest that anyone who thinks that Americans are some of the world's worst polluters are not very widely traveled.

We are declared to be *polluters* because we exhaust vast quantities of carbon dioxide from the smokestacks of our power plants. The rest of the world uses electricity, too, so they are doing the same thing. But life in third-world countries is very different from life in America.

The author has spent lots of time traveling in third world countries. During those travels, practically every large truck belched copious amounts of black smoke and every automobile and motorcycle was slightly out of tune. The roadways were generally polluted to the point that it was difficult to breathe and the roadways were also generally dirty with grayish black dust. To roll down automobile windows and deal with the fumes directly was quite nasty. To walk to market alongside those same roads was even worse.

The waterways in third world countries are equally polluted compared to ours. Lots of sewage flows directly into their waterways which are also then used to launder their clothes. It's absolutely amazing to see. And then, of course, there is the supply of drinking water. It comes from the same sources, too.

The author has also visited several of America's big cities, and they are very clean by comparison. Sure, we have pollution to deal with, but nothing like I've seen in countries around the world.

Yet we are told constantly that we are the world's polluters. And much of that is because of our carbon dioxide production. In my opinion, such complaints are all baloney! Carbon dioxide is a naturally formed gas that is crucial to life on this planet. To label it a pollutant gas, as our Supreme Court did recently, is absurd! To complain about it, label it a major pollution problem, and begin to formulate regulations to reduce it is just nuts!

To create tons of carbon dioxide from the combustion of fossilized vegetation and return it to the atmosphere from whence it came is not my idea of pollution. It is the completion of its natural cycle. Only through the use of gaseous carbon dioxide can solid forms of carbon be formed again to create tons and tons of new solid carbonaceous (organic) plants. Only in the form of gaseous carbon dioxide can fossil fuels complete their natural cycle that started many thousands of years ago.

To jump in and artificially alter this natural cycle for reasons of global warming could, in fact, negatively and irrevocably affect the earth's natural oxygen/carbon dioxide cycle. And if God is actually controlling the earth, the atmosphere, and the wind and the rain (see Chapters 3 and 4), new legislation may directly fight against His wishes by attempting to alter this balance. Do we really know what we are doing to try to legislate such changes? Have we thought this out completely? I don't think so!

The Atmosphere & Greenhouse Gases

In the previous chapters, we mentioned our current global warming problem (the one we're told we have) —- which is supposedly due to increased carbon dioxide concentrations in our atmosphere. Let's consider this in a little more detail by studying the composition and properties of our atmosphere.

Composition of Air

The volume and weight percentage compositions of earth's atmosphere are shown in Table 9.1:

Table 9.1 Composition of Earth's Atmosphere at 60°F

	Volume %		Weight %	
	0%RH	100%RH	0%RH	100%RH
Oxygen	**20.99 %**	20.623 %	23.20 %	22.944 %
Nitrogen	**78.03 %**	76.665 %	75.46 %	74.631 %
Argon	**0.94 %**	0.924 %	1.30 %	1.285 %
Other	**0.04 %**	0.039 %	0.04 %	0.045 %
Water		1.749 %		1.096 %

The two columns for each basis show the composition as water varies from 0% to 100% relative humidity.[198] The values listed in the first column (bold type) are the most commonly reported values.

The "Other" gases in this chart include carbon dioxide, hydrogen, neon, helium, krypton, and xenon. When the North American Combustion Handbook was published in 1978, carbon dioxide, comprised about 0.03% by volume of the air.[199] Currently, carbon dioxide in the atmosphere is reported to be about 0.039%[200] and all other percentages are just slightly smaller.

Heat Contents of Air Gases

The heat contents of common gases in BTU/scf (British Thermal Units/standard cubic foot) and BTU/lb at 100°F relative to 60°F are shown in Table 9.2.

Table 9.2 Heat Contents of Atmospheric Constituents at 100°F, relative to 60°F[200]

	BTU/scf	BTU/lb
Air	0.73	9.6
Carbon Dioxide	0.96	8.3
Hydrogen	0.73	139.1
Water Vapor	0.84	17.7
Nitrogen	0.73	9.8
Oxygen	0.74	8.7
Liquid Water		40.0

This table shows relative numbers[201] based on the heat content in each gas at 60°F. For instance, to raise the **air** temperature from 60°F to 100°F, one

---

[198] North American Combustion Handbook, North American Manufacturing Company, Cleveland, OH, 1978, p. 1.

[199] *ibid.*

[200] Wikipedia, "Carbon Dioxide."

[201] North American Combustion Handbook, *loc.cit.*, p. 58.

would have to add 0.73BTU to each scf of air. Another way to say this is that each standard cubic foot of air at 100°F holds 0.73BTU more heat than one standard cubic foot of 60°F air.

The values shown in Table 9.2 for water vapor do not contain latent heats. Latent heat must be taken into account for water because water can change phases from liquid to vapor in this temperature range. None of the other gases change phases within the normal range of climate temperatures. Water is unique in this regard.

Note that one standard cubic foot of 100°F carbon dioxide holds only slightly more heat than does one standard cubic foot of 100°F air. Water vapor at 100°F also holds only slightly more heat than air. The heat contents of the other gases are about the same as air. Overall, however, all of the gases hold similar amounts of heat in this temperature range.

Latent Heats

Because water can undergo two possible phase changes within normal temperature ranges, we must consider latent heats.

The latent heat of a substance is the amount of heat required for that substance to undergo a phase change at constant temperature. Ice may warm up to 32°F, but it requires additional heat to convert it to liquid water at 32°F. The amount of heat required is known as the *latent heat of fusion* of water. Liquid water also can be converted to water vapor at a variety of temperatures. To do so, additional heat must be added to convert the liquid water to water vapor. The amount of heat required in this case is known as the *latent heat of vaporization* of water.

A common example should help our understanding of this quantity known as latent heat. Consider a pot of water on a kitchen stove. Adding heat to it will raise the water temperature to its boiling point at 212°F. At that point, further heat added to the pot will not raise the temperature of the liquid water any further. The liquid will remain at 212°F, but the water will be bubbling. The bubbles mean the water is boiling. All of the heat added when the water is boiling causes liquid water at 212°F to convert to water vapor (steam) at 212°F. As long as there is water in the pot while the pot sits on a hot burner, the temperature of the boiling water will remain at 212°F.

If the pot has a lid, which defines a closed container, when all the liquid has been converted to steam, the addition of more heat will cause the temperature of the steam in the pot to rise above 212°F. Without lids, pots of water are open containers. In an open container, once the liquid converts to vapor, it will disperse throughout the atmosphere in the room and move away from the pot. When all liquid is finally converted to steam from an open pot, the pot will be empty. Further additions of heat at that point will cause only the metal of the pot to heat beyond 212°F.

Now let's consider the actual values for latent heat. The latent heat of vaporization for water at 100°F is 1037.0 BTU/lb.[202] The latent heat of fusion for water at 32°F is 143.5 BTU/lb.[203]

This means it takes 143.5 BTU to thaw a pound of 32°F ice to produce 32°F liquid water, and it takes 1037.0 BTU to vaporize a pound of liquid water at 100°F to form water vapor at 100°F.

Air Heating Example

Now let's go back to the subject of interest. How much energy is required to heat air, carbon dioxide, and water from 60°F to 100°F?

To heat 1 lb of air from 60°F to 100°F requires 9.6 BTU (Table 9.2). To heat 1 lb of gaseous carbon dioxide from 60°F to 100°F requires 8.3 BTU. To heat 1 lb of liquid water from 60°F to 100°F requires 40.0 BTU. But what if the water vaporizes? To heat 1 lb of liquid water from 60°F to 100°F and then to vaporize it to form water vapor at 100°F requires 1077.0 BTU (40.0 BTU to heat the liquid water to 100°F + 1037.0 BTU to vaporize it at 100°F.)

Notice that when the possibility of vaporization of water is included in the calculation, water absorbs about 130 times as much energy as is required to heat carbon dioxide over that same temperature range.

Considering these values, which of the gases or water absorbs the most solar energy as it heats from 60°F to 100°F? 9.6 BTU, 8.3 BTU, or 1077 BTU??? How much liquid water is available to absorb the sun's rays? Considering that most of the earth's surface is covered by liquid water, there is an almost unlimited supply of water available to this reaction.

This factor in the global warming debate is not usually emphasized: **Water vapor** is **THE most prominent green house gas** in the earth's atmosphere.

---

[202] *ibid.*, p.348.

[203] Wikipedia, "Enthalpy of Fusion."

Greenhouse Gases

Greenhouse gases, by definition, are those gases that can easily absorb infrared radiation from sunlight. Of the most prominent gases in our atmosphere, water and carbon dioxide are the most abundant greenhouse gases. Methane, nitrous oxide, ozone, and chlorofluorocarbons[204] are other greenhouse gases present in the atmosphere at much smaller concentrations. All greenhouse gases are known to absorb and re-radiate infrared solar radiation.

Incoming solar radiation travels directly in line with the sun, but scattered or re-radiated photons can travel in all directions. This means that greenhouse gases can absorb and re-radiate the sun's radiation directly or they can absorb it from re-radiated or scattered sources.

Consider the vast quantity of liquid water at the earth's surface. Sunlight plus water produces both warmer water and warmer water vapor. The conversion of liquid water to water vapor accounts for the absorption of huge quantities of radiant solar energy. The water to water vapor phase change is by far the biggest absorber of the sun's energy in our atmosphere.

Next, let's consider the amount of carbon dioxide present in the atmosphere, compared to the vast quantities of surface water, airborne water (cloud and rain droplets), and water vapor on the earth and in its atmosphere.

Concentration of Carbon Dioxide in Air

Relative to the preceding discussion, and to the broader global warming discussion, the percentage of carbon dioxide in the air needs to be considered. It is currently about 0.039% by volume. That is 3.9 molecules of carbon dioxide exist for every 10,000 gas molecules that make up the air.

Turn back to Chapter 4. It contains approximately 50,000 characters. Make believe each of those characters is a gas molecule in the air we breathe. Twenty of those characters are bold gray script, like THIS, to simulate carbon dioxide molecules. The special characters in that chapter are a picture of the concentration of carbon dioxide molecules in the air we breathe. That is the amount of carbon dioxide we are being told is going to lead to catastrophic warming on earth years in our future.

---

[204] Wikipedia, "Greenhouse Gases."

Actually, they predict that the percentage of carbon dioxide in our atmosphere is going to double in coming years. That means 8 molecules out of 10,000, or double the number of special characters in the whole of Chapter 4, are going to cause the global warming problems. Does that make sense? Only a very few molecules of carbon dioxide exist among the very many molecules that comprise the air.

The total number of gas molecules in air is a huge number. Even with only 0.039% concentration of carbon dioxide in air, plenty of carbon dioxide molecules are present in each volume of air. For the atmosphere to heat, however, the energy gained by each 4 carbon dioxide molecules must be evenly distributed throughout the remaining ~10,000 other molecules. Is it plausible that 4 molecules out of 10,000 can have such a substantial effect on air temperatures? It would make more sense that the 21% oxygen in our atmosphere would cause problems, or the 78% nitrogen would, but the dire warnings are being waved due to the 0.039% carbon dioxide — while the biggest absorbing molecule, the water molecule, is ignored. It would be more believable if water were blamed. But even the alarmists know that nothing can be done about water and water vapor in our world. That leaves the next most abundant greenhouse gas, the 0.039% carbon dioxide, to take the blame.

And a lot of that blame falls on the combustion of fossil fuels, which everyone knows is bad! ... so they say! They need a scapegoat in this story and carbon dioxide has filled that role.

The point of this discussion is to show that normal air only contains 4 molecules of carbon dioxide out of each 10,000 gas molecules. Those 4 molecules cannot be producing the enormous effects for which they are blamed, because the effects are limited by their low concentration.

Liquid water and water vapor have an enormous effect on air properties because the latent heat of vaporization of the water can absorb an enormous amount of energy compared to the other molecules. John Tyndall "concluded that water vapour is the strongest absorber of radiant heat in the atmosphere and is the principal gas controlling air temperature. Absorption by the bulk of the other gases is negligible."[205] How recently did he draw this conclusion? 1863.

---

[205] Wikipedia, "John Tyndall."

And now the US Supreme Court has labeled carbon dioxide a pollutant gas. Amazing! You would think carbon dioxide has no redeeming qualities at all.

Benefits of Carbon Dioxide

We mentioned that carbon dioxide and oxygen are in a natural balance within our atmosphere. Oxygen is at a much higher concentration than carbon dioxide, but the 21% to 0.04% ratio works well. Human and animal life rely on oxygen in the atmosphere; plant life relies on carbon dioxide in the atmosphere; and each seems to have enough of the required gas for life to continue well at these values.

Without sufficient plant life, or sufficient human and animal life to maintain the proper balance of these gases in our atmosphere, all life could disappear from this earth! ... or life could become much more difficult to sustain.

Imagine an earth with no carbon dioxide in its atmosphere. Photosynthesis would not be possible, so there would be no plants, no food sources, nor a ready source of oxygen for humans and animals. Or imagine an earth with no oxygen. Human and animal life would not be possible and there would be no ready source of carbon dioxide for the plants.

But the two gases are in relative balance with one another; they have existed at approximately this balance (we think) for millennia; and life as we know it, among humans, animals, and plants is continuing very well.

Studies have shown that plants grow faster when concentrations of carbon dioxide are elevated beyond current values. Plants grow up to 50 percent faster when carbon dioxide concentrations in our atmosphere are higher — as much as 1,000 ppm CO_2.[206] The current value, 0.039%, is 390 ppm (parts per million). If the concentration of carbon dioxide in the air increases to a little more than double its current concentration, that would help plants grow faster. Other studies have shown that warmer temperatures also favor faster growth of plant life.

---

[206] Wikipedia, "Carbon Dioxide."

So because someone wants to label fossil fuels in general, and oil in particular, to be bad for the planet[207], we are seeing carbon dioxide used by politicos to change the economic conditions of our country. Why haven't we seen even-handed presentations on this subject in the mass media?

Some might say carbon dioxide is the culprit because they don't understand the argument. I maintain that they do understand! They have a political agenda which overrides common sense as it requires them to demonize carbon dioxide. Consider it demonized!

Summary

Carbon dioxide is a natural part of earth's atmosphere — but it is a very small part (0.039%) of it. Carbon dioxide is also a greenhouse gas, an absorber of infrared radiant energy. One would expect carbon dioxide to not be a major issue because it is present in minute concentration in our atmosphere. Water, on the other hand, is a great absorber of infrared energy. If the radiant energy is not absorbed by the water vapor in our atmosphere, it has an excellent chance of being absorbed by the liquid water that covers most of the face of the earth. Water vapor, like carbon dioxide, can absorb only a little energy. Liquid water can absorb a lot!

The latent heat of vaporization of water is a much greater absorber of energy than the dry heat absorptions of the gas molecules that form our atmosphere. The only atmospheric constituent that has a phase change within the normal range of earth's temperatures is water — and it has two phase changes, not just one in that temperature range.

Since the latent heat of vaporization absorption by water is more than 100 times greater in magnitude than the absorption energies needed to raise gas temperatures, this author agrees that Tyndall's conclusion was correct: Water vapor is "the principal gas controlling air temperature."[208] Carbon dioxide is not the guilty party.

These two reasons (low carbon dioxide percentage, and large latent heat of vaporization of water) suggest that carbon dioxide is not the perpetrator

---

[207] Bush, G.W., President, 31 Jan 2006, State of the Union Message, "America is addicted to oil."

[208] Wikipedia, "John Tyndall."

of global warming. In fact, instead of arguing about carbon dioxide as **the** cause for global warming, we should return to the discussion to determine whether or not global warming currently exists.

The other discussion we could have is whether or not mankind should be making any attempt at all to try to control (or alter) earth's temperatures. The verse from Proverbs 16:25 seems to apply: "**There is a way that seemeth right unto a man, *but the end thereof are the ways of death.***" Does mankind really know enough to attempt to alter earth's climate? I don't think so.

11

The Big Bang & Einstein's Equation

Many Christians find *The Big Bang Theory* insulting and incompatible with their belief that God created the heavens and the earth. Many scientists feel the same way towards the Christians' belief in the Creator God.

In this short chapter, I'd like to make two simple observations. One concerns the Big Bang, and the other concerns Einstein's Equation.

Hopefully, each will provoke thought on the subject of how the earth was created.

The Big Bang

The Big Bang Theory is a recent attempt by mankind to explain how the universe was created. Physicists and other scientists want to explain the earth's creation in logical, physical terms. They don't want any miracles or other religious artifacts to cloud their explanations. Everything that occurred, they believe, should be explainable with natural phenomena. The thinking is that mankind should certainly be capable of understanding and explaining the details of creation using the known laws of physics.

On this point, as well as on the subject of global warming, many of our scientist friends have taken arrogant stands similar to the one Job took for which he was confronted by God. Job thought he knew God and could explain how God operated. But he could not. And when confronted by God about it, there was nothing he could say. So he remained quiet. He couldn't explain any of what God had done, nor could he explain why or how God had done it.

Some day, today's scientists may be confronted by God with the same words He used to confront Job. Then, they will learn that they, too, can say nothing. They, too, can listen to Him — and remain silent.

But until that point in time, they are doing their best to explain how the creation occurred apart from God's help or control. This is why I label it *arrogance.* They are certain they can understand and explain everything about creation. It will just take them time to work it out.

Most of their explanations to date are beyond the understanding of the average man. I can understand their broad strokes, but I really don't understand the precise details of their explanations.

The Bible, however, says that mankind does not understand God, or how or why God does anything — except by faith. **"There is none that understandeth, there is none that seeketh after God."**[209] **"O the depth of the riches both of the wisdom and knowledge of God! How unsearchable are His judgments, and His ways past finding out! For who hath known the mind of the Lord? Or who hath been His counsellor?"**[210] **"Through faith we understand that the worlds were framed by the word of God, so that things which are seen were not made of things which do appear."**[211]

Many men and women claim they have understanding. And according to them, a proper scientific explanation of anything precludes the use of God as a participant or a cause. So all of their attempted explanations make no mention of God whatsoever. By definition, God has no role in any scientific explanation of anything.

All that said, consider this: Whether or not you believe in the Big Bang Theory, it has a major point of agreement with the Bible. At some time in the past, according to the Big Bang theory, there was an instant — BANG!!! — when all was created.

The Bible, in Genesis 1.1, says, **"In the beginning God created the heavens and the earth."** Let me paraphrase that to make my case: In the

---

[209] Romans 3:11

[210] Romans 11:33

[211] Hebrews 11:3

beginning, God spoke — BANG!!! — and the heavens and the earth were created.

The God of the Bible is an eternal God. Consider a verse out of each testament of the Bible: "**The eternal God is thy refuge.**"[212] "**For the invisible things of Him from the creation of the world are clearly seen ... even His eternal power and Godhead.**"[213] God, His power, and His rank as Lord, are eternal.

But the heavens and the earth are not eternal. They had a beginning.[214] Going forward, the heavens and the earth may be eternal, but in the past — they are not.

Aren't the Big Bang Theory and the Bible in agreement on that point? The heavens and the earth each had a beginning — and they were both created in an instant (according to Genesis 1:1 and according to the Big Bang Theory)?

Everything as we know it today, including life on this earth, didn't form in that one instant. The Bible explains the creation over seven days. Physicists explain that the Big Bang occurred and then everything developed. Their explanations of how life was created on this planet says that it formed over millions of years following the Big Bang by gradual, evolutionary processes. There is no agreement on the details.

But on the original fact of creation — BANG!!! — there seems to be agreement. And until the physicists and other scientists can believably fill in the rest of the blanks of creation, their explanation agrees with Genesis 1.1. Think about it.

I think Christians should stop getting upset with the explanations coming from today's crop of physicists. Agree with them. Point out that they have come to the same conclusion the Bible makes in its first verse. See how they react. See what they have to say, about that!

---

[212] Deuteronomy 33:27

[213] Romans 1:20

[214] Genesis 1:1

Just maintain a pleasant smile during the whole conversation. After all, they still haven't explained where all of the energy comes from to produce the Big Bang. On that point, they still face a dilemma.

Einstein's Equation

This equation proposes the equivalency of matter and energy.

Soon after the turn of the 20th Century, Albert Einstein suggested an equation for this phenomenon:

$$E = mc^2 \quad (1)$$

where E = energy, m = mass, and c = the speed of light.

Most people are familiar with this equation. Energy, on the left side of the equation, is equivalent to a mass term and the square of the speed of light on the right side of the equation. Let's expand the equation a little bit by substitution, using the definition of speed as distance per time (x/t):

$$E = m(x/t)^2 \quad (2)$$

where E = energy
m = mass
x = distance
t = time
x/t = a speed — in this case, the speed of light, c

Equation 2 is the interesting form of the equation. The c in Equation 1 is a velocity — the speed of light. Velocity is distance per time. So when Equation 1 is expanded using *x/t* in place of *c*, it takes the form shown as Equation 2. Now, what does this new equation tell us?

God who is all energy (and no mass, space, or time), created the universe which contains all three: mass (m), space (x), and time (t).

In what time frame does God operate? A day with God is as a thousand years, and a thousand years is as a day.[215] It is not clear that God

---

[215] 2 Peter 3:8

throughout eternity past, or throughout eternity future, even experiences the phenomenon we know as **time**. We know that God is described by Jesus as "**a Spirit**" (who does not have a body).[216] But God represents enormous, unlimited energy! Imagine the vast amount of energy required to create the universe! Imagine the vast amount of energy dispersed at the Big Bang.

Now consider Equation 2. When God, who is timeless energy, created the earth, He gave it mass, He put the mass into space (a function of distance — that is, with length, width, and breadth), and He created time.

That is what Equation 2 says. Energy, on the left, is equivalent to a function of mass, space, and time on the right. He placed us here on the earth in a world of mass, space, and time.

Mankind thinks in terms of days, weeks, months, and years. But how does God think? How does Jesus think? How long, for example, did Jesus suffer on the cross? It all happened in one of our days. In fact, it all took place over about six of our hours.

In God's time frame, how long was that? When Jesus suffered in darkness on the cross for those 3 hours, how long was that from God's point of view? We don't know. It could have been an eternity. It probably was an eternity for Jesus.

So what is my point? Einstein's equation says that to come from God's realm (energy) into this world He created necessarily required Him to create mass, space, and time. If Einstein's equation is correct, that's what it shows: energy is equivalent to a function of mass, space, and time. He lives in His realm. We live in this realm. Good equation, Albert.

Give this some thought the next time you're resting. And pose this suggestion to your physicist friends next time you cross paths with them at a social gathering. Maybe this will become a new topic for discussion.

---

[216] John 4:24

12

The Global Warming Debate

It depends whom one listens to — global warming (GW) is established fact — or it is fiction ... a hoax! The fact side is prominently represented daily in the mass media (to include newspapers, cable news, radio, and TV). Those who believe it's a hoax are represented on talk radio and the internet.

Scientists are split over support for GW. Some are 100% supportive of the dire predictions — some are 100% against them. Most news outlets support both the alarmist claim that GW is fact *and* the political elites' suggestion that the debate is over. According to them, nothing remains to be decided — GW is fact.

Many industries are preparing for the next step — the sequestration of carbon dioxide. And lots of research is being funded by the government and various other research agencies. All is proceeding as if GW is fact.

One popular entertainment figure claimed recently that GW skeptics are no different than Holocaust deniers.[217] Throughout 2010, President Obama continued to push his "cap-and-trade" legislation designed to redistribute income from rich to poor corporations and from rich to poor countries by requiring limits on the production of carbon dioxide combustion products. If you exceed your government-imposed, (and almost surely to be) arbitrary limits, you will pay some company elsewhere for the privilege of using their carbon allotments they aren't using.

---

[217] www.foxnews.com, "McCartney, in Interview, Compares Global Warming Skeptics to Holocaust Deniers," 24 June 2010.

Ostensibly, "cap-and-trade" is a fix to the global warming problem. In actuality, it is a cover for a political redistribution of income agenda. President Obama wants to impose this legislation because he believes (at least publicly he says he does) global warming is fact!

Practically every time we have a hot day, members of the political elite and/or the media tell us that the heat is due to global warming.[218] Every time we have a big snow storm, like last winter's blizzard in the Northeast, they tell us the storms were caused by global warming.[219] Hurricane Katrina, we are told, was the result of global warming.[220] Every recent major weather event, we are told, was caused by global warming.

Recently however, Professor Phil Jones, the former Director of the University of East Anglia's Climatic Research Unit admitted that "there has been no 'statistically significant' warming over the past 15 years."[221] Why is this important? Professor Jones and his research unit have been in the center of the spotlight for a long time, saying that their data proves anthropogenic (man-made) global warming. Their unit was at the center of the original global warming fanaticism. Now, after being a direct cause of the brouhaha, he has admitted there has been no recent global warming. Essentially, his admission can be paraphrased, "It was all a hoax."

It no longer appears to be a scientific debate; it is now a political debate — and political debates don't just stop because facts were inaccurate, or because someone lied. Whom should we believe? ... scientists (like those from East Anglia who admit they have been lying and playing games with the data for years)? ... or the GW 'skeptics' (like this author)? Keep in mind that the same fanatics who believed East Anglia's original opinions now don't believe it matters what East Anglia has to say. Anyway, according to the alarmists, the

---

[218] Sheryl, "Is July 2010 Heat Wave a Sign of Global Warming?", 31 Jul 2010, thisgreenblog.com.

[219] Shibut, Denis, "Global Warming and The Perfect Snow Storm: Climate Change," 6 Mar 2010, WCW inSight, www.whitefieldconsulting.com.

[220] Environmental Defense Fund, "Hurricanes Stronger Due to Warming," 29 May 2007, http://www.edf.org/article.cfm?contentID=6452

[221] Petre, *loc.cit.*

whole debate is all the result of propaganda funded by the evil oil and coal companies.

There are political agendas at work in this debate. Those in favor of GW legislation and cap-and-trade appear to have two goals in mind: (1) in the name of saving the planet, they plan to ruin the US economy; and (2) many of GW's leading supporters personally stand to make big bucks out of cap-and-trade. The President appears to be in the first category — the former VP in the second.

So whom are we to believe? What have we learned in this book?

Biblical Story of Creation vs Darwinian Evolution

We started by looking at creation according to the Bible. Many in the scientific community ignore the Bible — just because it is the Bible. They insist that the only valid explanation of creation will come from science using only natural physical explanations for everything. To them, the Bible represents religion, so its explanations simply cannot be true.

But that is an invalid conclusion. **Just because modern science doesn't believe, or cannot corroborate, the Bible's explanation doesn't mean that the Bible is wrong.**

Modern science cannot corroborate the Darwinian theory of evolution either. But that hasn't stopped them from putting full faith in it. More and more scientists are openly questioning evolutionary theory — especially in light of the most recent details we have learned about life since the advent of the electron microscope. No one in Darwin's day (Darwin included) imagined the kinds of detailed, minute structures and machines that have been identified residing and functioning within living cells. Darwin and early followers expected to find only a simple viscous fluid (phlegm??) within each cell. But surprise! Surprise! They found more and more complexity of structure and function as they delved deeper and deeper with modern microscopes into living cells.

For years, we have heard arrogant statements by top evolutionists that they only needed to run a few *well-designed* experiments to conclusively prove the truths of evolution. Those scientists implied that with their mighty intellects, they could easily define the right experiments, and we would all see the veracity of evolutionary theory! Statements like that have come and gone — and we're still waiting. Some of those scientists have come and gone, too — and we're still waiting.

Evolutionary theory cannot be proven. They haven't been able to prove it yet, and they appear less and less likely to be able to do so. On the other hand, it is impossible to prove a negative. It is impossible to prove, conclusively, that macroevolution **does not exist.** The very next experiment could solve the puzzle. The fact that they haven't proven evolution yet doesn't mean that it won't happen. So we continue to wait for proof. . . . and wait . . . and wait . . . and wait.

The Biblical story cannot be explained scientifically either. Most of its major points fall into the "miracles" category and science and evolution supporters simply refuse to accept the miraculous as a valid answer to anything. That the Biblical story requires unexplainable miracles doesn't make it wrong. It just makes it "unacceptable" to many people. Simply put, many **refuse** to believe the Bible's explanation because it has to do with God. And most evolutionists want nothing to do with God, or with any creation argument that credits God.

What is so unbelievable about God? What does the Bible say? Here's what we learned.

God Created

According to the Bible, God spoke, and the worlds were created. He created the heavens and the earth and all life. Many people, even those who are unfamiliar with the Bible, realize that the "creation story" can be found in the first two chapters of the Bible.

But statements that "God created," are repeated over and over throughout the whole Bible. They occur numerous times. That God created was claimed and attested to by many writers of scripture.

Do the heavens and the earth need help as life proceeds? Yes. And that is the next point we considered.

God Maintains

According to the Bible, God not only created the heavens and the earth, but He maintains them. He holds them together with His power. God's hand can be found in the everyday functioning of every living creature on earth. This, too, is ignored by many because it deals with the supernatural. Mankind can't explain how God can affect anything — so anything miraculous becomes an object of disbelief.

What does all this have to do with global warming? The next point we studied showed that God is in control of the wind and the rain.

God Controls the Wind and Rain

Some of the most frequent references throughout the Bible are to the fact that God controls the wind and the rain. That is, God controls the weather systems on the earth. He controls the wind, rain, clouds, storms, lightning, thunder, snow, hail, sleet, etc. He also controls the shining of the sun in the absence of cloudy, nasty weather. And through His control of the weather, He controls whether plant life thrives or fails.

Now this facet of God's control **IS** important to the global warming debate. Scientists are trying to predict climate and weather conditions hundreds of years into the future. They are using known physical principles, recorded data, and their understanding of weather systems to do it.

But there is no way for them to predict what God will do at any given moment with the weather. They simply cannot know in advance. So even their best efforts to model weather systems will fall short if God is actually in charge.

A recent presentation by Cotton[222] suggested that the climate 10 to 50 years from now is simply not predictable. According to Cotton, there are too many unpredictable phenomena that affect our weather to possibly make accurate predictions over that time frame. He didn't even consider control from God. He was talking only about the unpredictability of natural phenomena.

Cotton was humbled by his own inability to make accurate weekly predictions of winds and weather for his university's gliding club. Based on his experience, he wondered how anyone is going to make accurate predictions of weather 10-50 years into the future? That is a good question. He concluded it cannot be done!

Next we asked questions concerning man's importance on earth.

---

[222] Cotton, W.R., Colorado State University, "Is climate really predictable on 10-50 year time table?", 20 Jul 2010, Powerpoint presentation.

Man Is Important

Is mankind, who resides on this earth, really important to it? The Bible says man is important — even special. Man was created in the image and likeness of God and given dominion over all life on the planet. I'd say that is quite special!

The Roman Catholic church, years ago, concluded that man and earth were so important, they made the geocentric model of the solar system into a church doctrine. Remember, the geocentric model puts the earth and mankind at the center of the universe. They were correct in their evaluation — mankind and earth are important. They had no business, however, turning that scientific theory into a church doctrine.

When Copernicus suggested, and Galileo taught, that the earth actually revolved around the sun (which it does), they considered those words heresy. Although Copernicus was making a scientific statement, the church understood him to be suggesting that **mankind and earth are not special.** That idea came to be known as the Copernican Principle. Copernicus died before the church could trouble him, but Galileo, who used his telescopes to confirm Copernicus' theory, spoke out just in time to be labeled a heretic.

Turns out that Copernicus and Galileo were correct. The earth and the rest of the planets do revolve around the sun. The earth does not reside at the center of the universe. But mankind and earth are very important parts of this creation! And in that regard, the church had the right idea.

Adam failed to fulfill all that man was supposed to do in this world, but Jesus fulfilled all of man's potential. And even though mankind now remains a fallen race, compliments of Adam and Eve, mankind is still very special in God's eyes. Mankind always has been special to God, and mankind always will be special to God!

How important are we? God designed this wonderful world in which we live "**to be inhabited**," and He designed mankind and all other life on this planet to occupy it. That's pretty important! That's pretty special!

Many proponents of uniformitarianism and Darwinian evolution, however, still subscribe to the Copernican Principle. They find nothing special about the earth or mankind, and any day now, they will identify other alien civilizations. We just happen to be here as the result of a long string of chance occurrences and mutations. God had nothing to do with it.

God's Intelligence Designed the World

Many today are trying to prove that intelligence was necessary to create the heavens, the earth, and all life. Biochemistry Professor Behe, at Lehigh University, uses the idea of "irreducible complexity" to question the veracity of evolutionary thought.[223] Many natural microscopic machines that function within human cells are just too complex to have gradually evolved to their current states. They are "irreducibly complex."

As science has been able to see deeper and deeper into living cells, we have learned that living cells contain very complex, microscopic machines made of numerous complex proteins. The more our microscopes were able to delve into the mysteries of cells, instead of finding uniform, unstructured goo residing in the cells, we found incredibly complex structures with equally complex functions. The finer we are able to see, the more detailed and complex designs we find.

Irreducible complexity suggests that intelligence was required to plan all of these details. It also means that at some microscopic level, a level of complexity exists that cannot have evolved in a step by step manner ala evolutionary theory. Two simple examples frequently used are the functions and structures of the eye and of the bombardier beetle's self-defense mechanism to evade predators. Too many integrated complex structures exist in those two examples for their functions to have evolved simply, step by step, consistent with Darwinian evolution. And those are just two examples of the myriads of complex structures in existence. Each irreducibly complex structure is a road block to a Darwinian explanation.

The conclusion is that intelligence was required to design and build such systems. Examination of these microscopic machines shows that an excellent engineering mind was at work during the design phases of all plants, animals, humans, and the earth.

As an engineer, the author knows the requirements for an excellent design. No man-made structures exist that compete in terms of complexity and elegance with the designs of the human body and the animal and plant life on this planet.

Take the most complex mechanical device you can imagine and compare it to the human body. There is **no** comparison. The human body can

---

[223] Behe, *loc.cit.*

perform a wide variety of functions, is elegantly complex, can reproduce itself, and has a wide variety of hidden functions that are just waiting to be called into use. Throughout a lifetime, some of those hidden functions may never be required — so they remain hidden and unused. But they are there! ... ready and able when and if needed.

Then there are functions that work in conjunction with the outside world. For example, how did the human body learn to sweat to cool off when sweating requires the cooling effect of the evaporation of moisture to produce cooling? It just happened by chance? How did happenstance learn that evaporation produces a cooling effect? If only we could teach future engineers to pay attention to such details in their designs!

Evolutionists would have us to believe that this was the result of sheer chance. As an engineering professor, I think that is baloney. This is a good example of a wonderful engineering design — and many such examples exist.

We should all be thrilled that we were made by God in His own image and likeness. But that is not good enough for many. I guess many are insulted by that suggestion. They would rather trust in the pure chance of evolution than to trust in a God who willingly put His knowledge, understanding, and power into His designs.

As the author's father-in-law used to say, "If a person wants to believe he or she evolved from a glob of snot floating on the ocean, that is okay with me." He put his trust in God and God's ability to design, engineer, and create elegantly complex, functioning systems.

Computer Models Are Simplifications

The Bible indicates that God controls our weather systems. His ways are unpredictable. Some scientists say that other climate change phenomena are also unpredictable. Climate change scientists, nevertheless, are attempting to create computer models to simulate weather systems and to predict climate conditions years into the future.

In an earlier chapter, we discussed whether all pertinent phenomena were included in these models, whether each phenomenon was properly represented by appropriate mathematical algorithms, and whether the models have been properly tested against current data. The best answer available seems to be: "we don't know." Why? Because the details necessary to answer all such questions are widely buried in the climate change technical literature.

We are constantly bombarded in the open mass media with claims that we are going to suffer dire consequences if we don't immediately change our ways. But the details of those studies and computer models have not been widely reported in the same, open mass media.

We simply need to go to the technical literature, I guess, and learn the answers. Why not? Because most such studies, results, and conclusions have been written for the climate science community who are fully literate in that subject. Most such papers are difficult to read and to understand — even by engineers and scientists who are technically literate, but not necessarily in climatology.

It makes one wonder who translates those results ... for the reporters who write articles for publication in the mass media? ... or for the political class who are constantly beating us over the head with dire global warming predictions? ... or for politicians who are developing legislation to combat global warming? ... or for celebrities who know nothing much about anything technical, but who continue to beat the drums in favor of global warming legislation.

Dr. Spencer, a Principal Research Scientist at University of Alabama Huntsville, has been saying for a long time that current models do not adequately account for cloud phenomena. Cloud cover and related phenomena must be properly simulated in good climate models. Are they being simulated completely and accurately? Dr. Spencer certainly doesn't think so.

Then, there are those who believe many phenomena that affect climate are not predictable **at all**. Professor Cotton[224], Professor of Atmospheric Science at Colorado State University, concluded that **it is not possible** to predict weather and climate conditions 10-50 years from now. How are the current models treating those phenomena he questioned?

That brings us back to one of our original questions: Are all phenomena adequately represented in today's computer models? It appears that the phenomena with the longest natural cycles (those that cycle over several decades or even centuries) are probably not adequately represented — if they have even been properly identified.

If one climatologist does not believe that all phenomena are adequately represented in today's models, and another questions whether or not it is even possible to properly simulate all required phenomena — why, then, are we discussing legislation to regulate greenhouse gases to reduce global warming?

---

[224] Cotton, *loc.cit.*

And if the Bible is correct and God actually does control the wind, rain, and weather systems, and that factor cannot be adequately represented in today's models — why are we moving forward so quickly?

Why aren't we discussing whether or not global warming is actually taking place? Why didn't the recent admission that there has been no global warming over the past 15 years automatically end the debate? Why aren't we discussing whether or not it's even possible to model future climates?

Self-Regulated Control Loops Are Very Capable Designs

We discussed process control loops in one chapter. Engineers design and implement such loops to automatically control processes. Actually, we confined the discussion to relatively simple control loops — those with only one variable to be controlled.

There are many natural, self-regulated control loops in existence in the world today. A self-regulated control loop is one that maintains the variable to be controlled without the need for the imposition of an external control system.

Some industrial systems, that are naturally self-regulating, still need good automatic control loops to work properly. Why? The original self-regulating nature of those systems may not regulate to the desired precision. But there is no way to add automatic controls to our climate system. It is beyond our abilities.

There is another reason why good self-regulating systems are difficult to design. Most self-regulating systems in nature have multiple controlling and controlled variables. Consider again our climate system, which is a gigantic, self-regulating system. Cloud cover, sunlight, humidity, surface materials, wind, vegetation, man-made objects, precipitation, etc., all affect air temperatures. Each of these also has an affect on wind. How would anyone try to control that system if they wanted to? We don't try to **control** the weather — we try to **predict** the weather. And we don't do a very good job of that!

In such a large, self-regulated system, all of the phenomena interact with one another. Their interactions produce relatively constant weather conditions. And those conditions happen to be in the range of livable temperatures.

Each phenomenon has opposing phenomena to keep it under control. Too much sun light could raise temperatures too high, but as surface temperatures increase, water vaporizes more easily, vapor drifts up into the atmosphere, clouds form, and the clouds block further sun light from reaching the surface. Too little sun light could have an equally damaging effect. But the atmosphere can hold only so much water vapor. Eventually, it rains or snows, the clouds disappear, and sunlight reaches the surface again. Many such interactions participate in the control of surface temperatures.

We know from our Army's experiences in Iraq that people can live in temperatures north of 120°F — it's not pleasant but it can be done. The author knows from experience that in winter, people can live in temperatures below -40°F. That's not pleasant either, but it can be done. These are two examples of extreme temperatures. Our self-regulating atmosphere tends to keep average temperatures at much more liveable values within that range.

We don't need to do anything and our temperatures, weather, and climate remain relatively constant. That is how a good self-regulating system works. It takes care of itself.

Try to design a self-regulating system and you will find that it is incredibly difficult to do. To design a self-regulating system on the scale of our climate — only an intelligent God could do that. Mankind is **not** smart enough to develop such a system. And the randomness of evolutionary theory has no smarts at all. It certainly cannot design such a complex system.

To design something as complex as the human body, which utilizes numerous automatic control loops, is beyond man's ability. To design something as complex as the whole universe is also beyond man's ability. Only an intelligent God could make such perfect designs. But we are supposed to trust (according to the evolutionists) that random chance could do all of that. Simple unbelievable!!

Natural Cycles

We also considered natural cycles that occur all over the earth. These, too, are examples of great engineering designs — especially when one considers the balances implemented by each cycle. The water cycle is an example of a very elegant cycle. Liquid water changes to water vapor, clouds, and rain — which is back again to liquid water. The water cycle includes lots of complexity plus lots of ways to affect the concentration of water vapor in the atmosphere.

The most important cycle on earth appears to be the natural balance between oxygen and carbon dioxide. Once again, the elegance of the balance is truly amazing — humans and animals use oxygen for life. They are perfectly balanced by the plants which use carbon dioxide for life. Humans and animals breathe oxygen and exhale carbon dioxide. Photosynthesis in plants uses carbon dioxide and releases oxygen back to the atmosphere. As long as the two remain in balance, this elegant cycle continues very well.

Carbon Dioxide in the Atmosphere

Since carbon dioxide gas and combustion of fossil fuels are the bad guys in this story (so we're told), we might wonder where all the carbon dioxide comes from? Obviously, human and animal life account for a lot of it. The burning of fossil fuels also accounts for substantial quantities.

Fossil fuels come from decayed plant life that has been compacted and coalified over many years. Fossil fuels start with plant life. The plant life grew originally using atmospheric carbon dioxide for photosynthesis. The burning of fossil fuels returns that carbon dioxide to the atmosphere, from whence it came years ago.

Further study also shows (and you don't hear much about this) that the increasing carbon dioxide in earth's atmosphere correlates very well with the earth's population.[225-226]

This brings us back to a point made earlier: as human population rises, increasing levels of carbon dioxide will help maintain the proper balance between human and animal life, and plant life. It is a very elegant natural cycle that maintains the balance between oxygen and carbon dioxide in our atmosphere.

In the name of solving the global warming problem, some people want to implement regulations that will inhibit this natural cycle and the oxygen/carbon dioxide atmospheric balance by removing carbon dioxide from flues before it can re-enter the atmosphere.

Removal processes could change the nature of the carbon dioxide from **organic** sources into **inorganic** products. Most inorganic forms of carbon dioxide are not easily converted back to atmospheric carbon dioxide for use by

---

[225] Cotton, *loc.cit.*

[226] Lewis, B.T.R., and the Ocean-499 class of 1997, "Atmospheric Carbon Dioxide as a proxy for growth of the human population?," U. Washington, School of Oceanography, 1997.

organic plant life. Once removed from this natural loop and converted to an inorganic carbon material, the carbon dioxide may no longer be readily available to help plant growth — and if plant growth is hindered, food sources may become more and more scarce.

That could produce disastrous results for society — results of the unanticipated kind. **And that is scary!**

Water Is Most Important Greenhouse Gas

Regarding warming of the earth, the most important greenhouse gas is water vapor. The process of converting liquid water to water vapor requires the absorption of an enormous amount of energy. When the water vapor condenses, that enormous amount of energy is given back to the atmosphere in the vicinity where the water condenses.

For this reason, liquid water is the most important form of water participating in this cycle. Its energy absorption capabilities simply dwarf those of gaseous carbon dioxide. The availability of liquid water to absorb sun light over the earth's surface effects a large influence on earth's temperatures. Six sevenths of the planet covered by liquid water versus ~400ppm carbon dioxide in the atmosphere is simply no comparison. Liquid water and the latent heat of vaporization dominate the heat absorption process.

The Big Bang Agrees In Part With The Bible's Creation Story

Rather than trying to shoot holes in the Big Bang theory, which most people don't understand anyway, it makes more sense to find the point of agreement between it and the Bible's explanation of creation. Both agree that the creation started in an instant. The physicists want to explain in great detail, step by step, following the Big Bang, using natural phenomena. They can't. The Bible explains that God spoke, and the worlds and all that live therein were created. Mankind refuses to believe it.

How did either of them actually happen? Science is trying to explain everything in great detail. Even if they could explain what happened following the Big Bang, they would then have to explain where all the energy came from that caused the big bang in the first place. The Bible, on the other hand, tells us that God created. How? We don't know exactly how it happened. But God certainly had the abundance of energy needed to create the universe.

We have lots of details of God's creation in Genesis 1 and 2, but that isn't sufficient for many. The author agrees with Job's answer, when he was confronted by God. "Were you there when I created? Explain the details if you can." Job was quiet. He said nothing. I agree with his decision. There's nothing **anyone** can say to God in answer to such a question.

God did it. **That** He created, and **that** He gave us really elegant designs is sufficient for all mankind to know. Further details are simply beyond our abilities to understand.

Energy Conversion to Mass Requires Space and Time

The other point I find extremely curious is Einstein's equation relating energy to matter. An eternal God who has the power to create the universe as we know it must have an incredible amount of energy available to Him. To create mass from that energy wasn't enough — He necessarily had to create all three: mass, space, and time. And here we are!! $E=mc^2$!

The Current Debate

How Were We Created?

Mankind arrogantly thinks they are as knowledgeable and as smart as God. When Job was directly confronted by God, he admitted he wasn't very smart — by keeping his mouth shut.

But that was Job. Many still believe they are smart enough to explain how the earth and the worlds were created. They are confident they can answer the questions that Job did not attempt. But rather than put their trust in an omniscient God, they have chosen to trust Darwinist evolutionary theory and random chance.

The evolutionary theory for the creation of life on this planet relies on randomness and chance. According to evolutionists, all of the magnificent designs represented by mankind, all of the different forms of animals, and the thousands of different forms of plant life formed totally without the help of intelligence! ... they formed by chance! All of the magnificent functions of the life on earth are mere chance designs that developed over millions of years.

In fact, Professor Richard Dawkins of Oxford University, who was described by Wieland[227] as "fanatically antitheistic" has stated that "Darwin made it possible to be an intellectually fulfilled atheist." Others have also made claims that evolution, taken to its extremes, leads to atheism. Randomness, chance, and the denial of any Godly intelligence, is completely compatible with atheistic belief.

So the two main arguments for creation are (1) God created, and (2) randomness and chance brought forth. Neither can be completely explained, but evidence shows that time, randomness, and chance lead to deterioration and decay — not to improved designs and elegance. A luxury car allowed to sit for a millennium will turn into a pile of dust. A pile of dust allowed to sit for a millennium won't turn into a luxury car — it will remain a pile of dust.

Global Warming? Predictable by Science?

Now here come the global warming catastrophists who can predict future climates. They must necessarily assume *uniformitarian* phenomena because anything else is unpredictable. Everything must continue to happen in the future as it always has happened in the past. That is *uniformitarianism*.

Cotton[228] recently concluded that future climates are unpredictable. Others, such as Immanuel Velikovsky in his 1950 book, Worlds in Collision, proposed *catastrophism* to explain how the earth formed. Some of the same evolutionists who staunchly opposed Velikovsky's catastrophist theories and who attempted to prevent the publication of his first book, have recently begun to propose their own catastrophist ideas for creation under the banner of *neo-catastrophism*.[229]

More recently, in their 2004 book The Privileged Planet, Gonzalez and Richards[230] wrote: "Most astronomers are convinced that a glancing collision

---

227 Wieland, Carl, "Dawkins and Eugenics — A leading high priest of evolution reveals its ugly side," 1 Dec 2006, http://creation.com .

228 Cotton, *loc.cit.*

229 Parker, Gary, "Neo-catastrophism," Ch 3: The Fossil Evidence, in Creation: Facts of Life, www.answersingenesis.org.

230 Gonzalez, *op.cit.*, p. 340.

between the proto-Earth and a smaller planetary body is the best explanation for the Moon's origin." Sixty years ago, catastrophist ideas were unacceptable. Now they are believed by most. I'd say that ideas are rapidly changing!

But if catastrophism is truly the means by which the earth and its moon formed, how do climate modelers take the possibility of future catastrophes into account? They can't.

If God really created the heavens and the earth, as we have discussed, how do climate modelers take God into account? They can't.

If weather forecasters can't predict this afternoon's weather, how can they possibly believe they can predict our climate a century from now? They can't.

Only arrogance puts us where we are. Job was directly confronted by God many, many years ago. Everyone here today still has their own confrontation with God to look forward to.

The best available evidence suggests two conclusions: (1) catastrophic global warming doesn't exist; and (2) global climates decades and centuries from now **cannot** be predicted.

Is The Political Class Really Looking Out for Our Best Interests?

Enter the radical political elite who have jumped on the global warming band wagon. What motivates them? Ambition? Power? Concern for the best interests of the governed?

The current President has demonstrated that he thinks he is smarter than the average person. Now that he is in office, he is going to do things his way, and he will cram legislation down our throats if necessary. He is supported by the political left, the leaders of Congress, and the media. What is in it for them? Power. Control over our lives.

Today's political community is as arrogant as the scientists who insist they can demonstrate the veracity of evolutionary theories by just designing a "few good experiments." They promise all sorts of things. But nothing has happened yet. We're still waiting. They will spend us into a good economy. They are spending the money. We're still waiting for the recovery.

The political elite also demonstrate their arrogance by acting like they know better than the rest of us. We are the "small people" after all. What do we know? We need universal health care, but we don't realize it. England's and Canada's universal health care systems don't work — but no matter. We need cap-and-trade legislation, but we don't realize it. We need to spend trillions of dollars we don't have, to fix problems that were created by the very people now in charge of solving the problems. That makes no sense, but the government is doing it! That is political arrogance!

According to our leaders, we have a global warming problem. They say we aren't taking it seriously. True! Why not? Because it is a bogus global warming problem. Because data shows we haven't had global warming for 15 years. Nevertheless, it is being crammed down our throats by those who stand to make lots of money and gain lots more power if they have their way. Where is their technical expertise, or their technical experiences that qualify them to address this issue? They have none.

I am certainly not one to look highly on the necessity for sheets of paper from colleges and universities (diplomas, if you don't catch my drift) and letters behind your name. Each of my diplomas plus 25¢ used to buy me a cup of coffee. Under most circumstances, they didn't mean very much.

What value are they today? You need them in most circumstances. Despite my years of experience and understanding of computers and computer programming, I have no documentation (college transcripts or diplomas) that attest to the fact that I ever studied computers. Think I could get a job with a computer company today? Used to be that you could be an engineer without a college degree, or a professor without a PhD. Not today! You need the sheet of paper and the letters behind your name.

I have a friend who was a plant manager. He held practically every job possible as he worked his way up in through the ranks. He has wonderful experience. Think he can get a job today in another company? He doesn't have a college degree so his applications don't make it past the first automated computer cut of applicant résumés. Why? You need a college degree to be a manager! Without that sheet of paper, your résumé is automatically rejected.

So college diplomas are mandatory. Correct? Almost — but not quite! If you happen to be a high-ranking supporter of the latest PC fad, or a movie star, or you are good friends with a powerful political leader, or you yourself are a political leader — you don't need one. Then, your position qualifies you to speak out on anything and everything — no sheets of paper required. Even with a sheet of paper, you can be a total novice, and be elected to the Presidency.

So the elite tell us that if we aren't climate scientists (a group which constitutes a relatively small fraction of today's scientific community), we have no right to speak out on the subject of global climate change. Yet every member of the elite class (politician, entertainer, and political activist, etc.) who is so inclined, can voice his or her opinion in support of radical climate change activists — and the mass media will dutifully report those comments.

Why should we pay any attention to any of them? Well, they are well-known entertainers! They are the political elite! They are the activists! Some of the actors played doctors or scientists once in a movie, so they understand science or medicine. That qualifies them to speak out on technical issues. Some have no apparent technical background at all. Many of these people are politicians and many politicians are lawyers. They have little if any training in the sciences. But since they have been elected, or they are popular, or just noisy, their opinions supposedly count!

Political Agendas

Everyone in Washington, so it appears, has their particular political agendas to further. Elected leaders have the best chances of seeing their pet agendas enacted. Not all such agendas are good, logical, or beneficial to the citizenry. That doesn't matter. Leaders will make the most of their power while they have it. The majority of Americans had absolutely nothing to say about the election of the Majority Leader of the Senate or the Speaker of the House, yet look at the power they wield! Are they watching out for our best interests? It surely doesn't appear that they are!

Current rumors are rippling about through cyberspace and on some political programs that many current Congressmen and Congresswomen who are expected to be voted out of office in November will return to Washington in a lame duck session in November and December to push through a radical leftist agenda before they actually leave office. Is that watching out for our best interests? That would be a definitive statement that they think they know more than we do — especially if they are voted out of office.

If that actually happens, what can citizens do about it? Vote them out of office twice? If they enact radical legislation in a lame duck session, citizens won't be able to stop them. We will just have to live with the consequences. If this happens, it will be an excellent demonstration that they are in it only for the power!

Regarding global warming, the cap-and-trade bill if passed will wreak economic havoc on this country's economy. That appears to be okay with current leaders. It will cause electricity rates to sky rocket. That's okay. The

President said so![231] It will also transfer wealth from the rich USA to poor countries around the world. That's okay, too!

This is a political agenda at work if I ever saw one. How do they accomplish it? In the case of global warming, they plan to prevent any further discussion of the issue. Don't let the 'skeptics' speak! Declare the discussion ended by **consensus!**[232] Science isn't decided by consensus, but in this case, they are trying to make the case that it is — and that the discussion is over.

Their benevolent goals in this matter have nothing to do with global warming. They have to do with power — and who holds that power. It has to do with the ability to tell Americans how to live and how to spend their money. It has to do with redistribution of income — from evil, rich capitalists to the poor. It has to do with ruining the capitalist US economy and rebuilding it in socialist form. It has to do with ignoring God and religious arguments, in favor of the brilliant intellects of our scientists and political leaders. It has to do with ignoring history — and then repeating it — to the detriment of the country. And we like sheep are supposed to quietly go along with everything.

Dirty Tricks

The fact that the mass media is almost totally supportive of this radical leftist agenda, sets the tone for any and all discussions. No offense to Senator McCain, but he was elected as the Republican candidate for President in the last election cycle with the help of the mainstream media. They announced which Republican candidate was in the lead, which candidate everyone should like, whom we should vote for, etc. It wasn't clear to me that McCain was actually in the lead until the press announced he was the front runner. "Look everybody — the others who have already voted have chosen him. Vote for him as well!" The media weren't accurately reporting news. They were attempting to mold public opinion. And they succeeded. Then, once the primaries were over, they switched their allegiance to his opponent. Once he was the Republican nominee for President, they had nothing good to say about him.

---

[231] Picket, Kerry, "Obama: Energy Prices Will Skyrocket Under My Cap and Trade Plan," 03 Nov 2008, NewsBusters.org.

[232] Wikipedia, "Climate Change Consensus."

The media also helped Senator Obama by not doing their homework. They cast him as the greatest Democrat to come down the pike, when he was really the least experienced of all candidates. People liked him because he was a good speaker. Many blacks voted for him simply because he is black. Many whites voted for him simply because he is black. What about his record? He had none to speak of. He was a community organizer. He had no management experience.

Now he's President and all citizens are learning that he has no leadership experience. But during the run up to the election — none of that mattered. What were his political leanings? He was extremely liberal — the most liberal senator in Congress. His voting record was available. But the media portrayed him as a centrist. Now, he not only has shown his liberal stripes, but he appears to be a socialist who doesn't even like the American capitalist system. Do you have a problem? Never fear — government will take care of it for you.

Everything that has made our country great is apparently bad in his eyes. Get rid of the free enterprise system. Replace it with big government. Let the government take over the automobile companies. Let the government take over the banks. Let the government take over health care. Force Americans to purchase health insurance. Get rid of free choice. The government can tell everyone what to do and what to think.

That whole episode — since the campaign began several years before the last election — until today — has been a gigantic dirty trick foisted on us by the political elite and the mass media. The American citizenry were duped by it.

The American people had the wool pulled over their eyes regarding the qualifications of our current leader. Now, the political elite and the media are attempting to do the same thing regarding global warming. We citizens don't know anything — they know everything. So we are supposed to just keep quiet and continue marching with the rest of the herd in whatever direction they point us.

Don't do it! Stand up and be heard!

Global Climate Change Conclusions

Global warming hasn't existed during the last 15 years. The global warming fanatics have, nevertheless, continued to push the coming dire consequences of anthropogenic global warming throughout those years. The constant drum beat: not only is the earth warming — but we're responsible for it. It's man-made!

We were told that global warming was responsible for any and every unusual weather event during that time. We still hear that. It is constant. Yet, we have had no global warming lately. We have gone 15 years without the left ever acknowledging there has been no global warming. It feels like global warming, so that makes it so!

The real culprit all this while has been the concentration of carbon dioxide in the atmosphere. It has increased a few thousandths of a percent over the last several decades. And earth's average temperature has increased a whole fraction of a degree! The sky is falling! The sky is falling!

What caused it? They say that carbon dioxide is the greenhouse gas that caused the problem. They even managed to have carbon dioxide labeled a pollutant gas by the US Supreme Court — so it must be a really bad gas!

Never mind that carbon dioxide and oxygen are at the center of the greatest, most important natural cycle on earth. Human and animal life depend on oxygen in the atmosphere. Plant life depends on carbon dioxide in the atmosphere. Each group exhausts the gas needed by the other — and so far, the two gases remain in relative balance.

The leaders of the free world want to put legislation into effect that may adversely affect this elegant balance. Why? Because our temperature has increased by a fraction of a degree over the last century. And we have this global warming problem (which doesn't actually exist.) Wink, wink!

In the name of global warming, mankind may now actually be about to do something that could really damage the natural balance between plant, human, and animal life.

The Bible, however, says God created. Mankind does not understand God's ways. We are governed by arrogant men and women who believe they know better than everyone else. And in the name of saving the planet from overheating, they may just stumble onto some regulations and procedures that will actually do harm to the planet.

The Bible says God not only created, but God maintains everything. God is in control. There is no doubt whatsoever that if mankind throws some monkey wrenches into God's natural designs — God can still handle it! But is that necessary? Must we arrogantly cause harm to the planet in the name of saving the planet?

It is time that men and women begin to ask questions like: Who created the heavens and the earth? Who is the God of the universe? Who controls the wind, rain, and weather? Then they should try to learn who God really is and what He can do. Certainly, and most importantly, they should stop listening to those who discredit the Bible just because it is the Bible, just because it requires faith, and just because it relies on the miraculous.

Just because **the Bible** says God created does not make that an incorrect statement. Just because the Bible is a religious document does not automatically mean its descriptions and declarations are wrong. Science has tried its best to explain creation and has to date been unsuccessful. Yet all the while, science ignores as fairy tales the Biblical explanations that actually do make sense.

Mankind has been created in the image and likeness of God. We are very special in God's eyes. If you want to believe evolutionary arguments which lead to atheism, be my guest. I choose to believe the Bible. I choose to believe the God who both created and maintains the world. I cannot explain everything He has done, but I trust that He knows what He is doing.

I hope you all come to that same realization, too.

The Author ...

Dennis Dinger is a Christian who is a Professor Emeritus of Ceramic and Materials Engineering at Clemson University. In 2008, he was disabled by a form of blood cancer. To date, chemotherapy and treatments have been successful. Over the past two decades, he has directed many applied ceramic engineering research projects, he has been an active researcher, private consultant, and the co-author and author of several books. He is co-author, with the late Professor James E. Funk, of the ceramic engineering textbook Predictive Process Control of Crowded Particulate Suspensions Applied to Ceramic Manufacturing, and author of three ceramic processing texts: Particle Calculations for Ceramists, Rheology for Ceramists, and Characterization Techniques for Ceramists. In addition to technical books, he is the author of The Coming of the Lord Draweth Nigh, a study of prophecy and the Revelation, The Tribulation to Come – A Study of the Revelation of John, and Absolute Truth for a Relative World.

www.ingramcontent.com/pod-product-compliance
Lightning Source LLC
Chambersburg PA
CBHW071430180526
45170CB00001B/288
* 9 7 8 0 5 5 7 8 0 4 4 0 5 *